Power Technology

Theodore B. Sauselein, P.E.

BNP Business News Publishing Company
Troy • Michigan

Copyright © 1994
Business News Publishing Company

All rights reserved. Except as permitted under the United States Copyright Act of 1976, no part of this publication may be reproduced or distributed in any form or means, or stored in a database or retrieval system, without the prior written permission of the publisher, Business News Publishing Company.

Library of Congress Cataloging in Publication Data

Sauselein, Theodore B.
   Power technology / Theodore Sauselein.
       p. cm.
   ISBN 0-912524-85-5
    1. Power (Mechanics)   I. Title
TJ163.2.S28   1994   93-5631
621--dc20   CIP

Administrative Editor: Joanna Turpin
Technical Editor: Barbara A. Checket-Hanks
Art Director: Mark Leibold
Copy Editor: Carolyn Thompson

This book was written as a general guide. The author and publisher have neither liability nor can they be responsible to any person or entity for any misunderstanding, misuse, or misapplication that would cause loss or damage of any kind, including loss of rights, material, or personal injury, or alleged to be caused directly or indirectly by the information contained in this book.

Printed in the United States of America
7 6 5 4 3 2 1

DEDICATION

I would like to dedicate this book to my wife Gayle for her forbearance. The computer is set up in the sewing room, and when work is in progress, books and papers are scattered all over the place, which has severely tested her doctrine of neatness and order.

ABOUT THE AUTHOR

Theodore Sauselein, P.E., C.E.M., earned his B.S. degree in electrical engineering from the University of Maryland. He is a registered professional engineer and is also a registered stationary and refrigeration engineer. Mr. Sauselein also holds certification by the Association of Energy Engineers as an energy manager. Mr. Sauselein's work experience has encompassed a wide range of industries, from ship building with Bethlehem Steel to his current position as a senior facilities engineer with AT&T Bell Laboratories. Mr. Sauselein is also the author of *Stationary Engineering for Boiler Operators*, which is a Business News Publishing Company publication.

TABLE OF CONTENTS

Chapter 1
ALTERNATING CURRENT 1

Chapter 2
TRANSFORMERS AND AUTOTRANSFORMERS 7

Chapter 3
SINGLE- AND THREE-PHASE POWER.. 13

Chapter 4
HAZARD PREVENTION AND SAFETY... 21

Chapter 5
DEMAND AND THE ELECTRIC BILL 29

Chapter 6
POWER FACTOR 35

Chapter 7
HARMONICS 45

Chapter 8
POWER LINE DISTURBANCES 51

Chapter 9
  LIGHTING TERMINOLOGY, SOURCES, AND BALLASTS 57

Chapter 10
  MOTORS: THREE-PHASE, SINGLE-PHASE, AND UNIVERSAL 65

Chapter 11
  MAINTENANCE 75

Chapter 12
  ELECTRIC METERS 83

Chapter 13
  ENERGY MANAGEMENT 93

Chapter 14
  THERMAL STORAGE 101

Chapter 15
  SYNCHRONOUS GENERATORS, INDUCTION GENERATORS,
  AND THE REFRIGERATION CYCLE 109

Chapter 16
  COGENERATION: WHAT IT IS, HOW IT WORKS,
  HOW TO USE IT 117

Glossary 123

Chapter 1

ALTERNATING CURRENT

VOLTAGE AND CURRENT: COUSINS TO PRESSURE AND FLOW

In contrast to what many people think, electricity is not a difficult process to master. It can, in many ways, be compared to the processes in hydraulic piping systems. In fact, the terms used to describe each of these systems are very closely related.

In order to gain a complete understanding of electricity and electrical systems we must first familiarize ourselves with the terms of the trade. The most common terms of measurement associated with electricity are voltage and current. These can be compared to the terms pressure and flow in piping systems. Table 1-1a further illustrates the relationship between the most common terms used in electrical and hydraulic systems.

A number of other comparisons can be drawn between electrical and hydraulic systems, as shown in Table 1-1b.

Now that we have familiarized ourselves with some basic terms and concepts regarding electricity, let's discuss alternating current and direct current.

Electrical system	Hydraulic system
Electromotive force (EMF) is the impetus that forces electrons through conductors.	Pressure is the impetus that forces water through pipes.
Volt is the unit of electromotive force.	Pounds per square inch (psi) is a common unit of pressure measurement.
Current is the amount of electrons flowing through a conductor.	Flow is the amount of water flowing through a pipe.
Ampere (amp) is the unit for measuring electron flow.	Gallons per minute (gpm) is a common unit of flow measurement.

Table 1-1a. *Comparison of electrical and hydraulic systems*

Electrical system	Hydraulic system
At a given amperage, more work can be done at higher voltage.	At a given amount of flow, more work can be done with higher pressure.
The higher the amperage, the greater the voltage drop.	The more flow in a pipe, the greater the pressure drop.
Large conductors handle high amperage better than small conductors.	Big pipes handle high flows better than small pipes.
Higher voltage requires thicker insulation.	High pressure requires pipes with thicker walls.
TVs don't last long when plugged into 480 volts.	It's hard to drink from a water fountain that's connected to a 1,000-psi source.

Table 1-1b. Comparison of electrical and hydraulic systems

WHY WE USE ALTERNATING CURRENT

First of all, **ac** stands for **alternating current**; **dc** stands for **direct current**.

Ac looks like a sine wave; it is defined as constantly changing in magnitude and periodically changing in polarity. Dc, on the other hand, maintains both constant magnitude and polarity. Batteries produce dc power.

So, why did ac win out over dc as the commercial power of choice? When Edison installed his dc-generating stations, he quickly realized that the power produced was limited to a short radius around the station; too much energy was lost as distances increased. To understand why, let's look at two basic power equations.

Equation 1:   *Equation 2:*
$P = V \times I$ $P = I^2 \times R$

Where: P = power in watts (W)
            V = the voltage
            I = the current in amperes
            R = the resistance in ohms

Note: *When equations appear in the book, don't worry. They are used only to make a point, not to train you for engineering.*

Equation 1 is the classic power equation. It shows that for a given amount of power, if the voltage is doubled, the amount of current required is halved. Equation 2 is used to calculate the power loss in conductors. (It is often referred to as the "I squared R" power loss equation.)

With the current squared in this equation, any reduction in current greatly reduces power loss. For example, if the current is reduced by half, power loss is reduced to one-fourth of its original value.

To put it another way: The weight of copper needed to transmit a given amount of power a given distance, with a fixed loss, varies inversely as the square of the transmission voltage. When transmission voltage is doubled, the weight of copper is quartered, all other factors being equal.

For example, at **100,000 volts**, the weight of copper required to transmit a given amount of power a given distance, with a fixed loss, is **one one-hundredth** of that required if **10,000 volts** were used.

This simply means that when electric power needs to be transported, it is best to do it at the highest practical voltage.

In Edison's time, 250 volts dc (vdc) was the highest level that could be used by commercial and residential customers. Because dc voltage can't easily be transformed to higher or lower levels, the generating station was limited to 250 vdc. Unless huge conductors are used, transmission losses quickly eat up 250 vdc.

Let's say that the nearest power station was about a mile from your house, and that 250-vdc, 100-amp service was required for your needs. For short runs, #1-ga wire with a diameter of 0.332 in. would be more than adequate.

However, a 5,000-ft run at 100 amps would experience a 160-vdc voltage drop; only 90 vdc would reach your house. Therefore, we need to use a larger conductor to lower the voltage drop. To get down to an acceptable 5% drop, a 1,250-kcmil conductor would need to be used.

How much copper does that represent? Well, each conductor in the pair of wires would be 1.29 in. dia, and the set would weigh around 1.9 tons. Let's say you own some stock in a copper company and decide to go ahead with the project. Now your neighbors want electric service. The existing wires can't be used — they're already at capacity. Each neighbor would need their own 1.9 tons of copper for service.

The tremendous advantage ac has over dc is that its voltage can easily be increased or decreased with transformers. The upper limit for ac generators is 23,000 volts ac (23,000 V or vac), but for efficient long-distance transmission, this is immediately increased up to 500,000 V.

Closer to the point of use, substations reduce this transmission voltage to a lower level for delivery to customers. Substations are a collection of equipment along a transmission line that control and switch electric power. Their transformers usually match the voltage to local requirements.

A popular voltage for residential distribution is 13,800 V. The power that you and your adjacent neighbors would need could be delivered on a pair wires the same size used for your telephones. A transformer near your house would then drop the voltage down to the level you require.

Industrial and large commercial installations are equipped to receive between 30,000 to 140,000 V. Residential distribution in the street is usually limited to 14,000 V, before being reduced to the 120/240-V level used in our homes.

Figure 1-1 shows the power house output voltage stepped up to the distribution grid voltage, then stepped down to match various customer requirements.

Another advantage ac has is that ac motors are simpler, cheaper, and require less maintenance than dc motors. When Tesla invented the three-phase induction motor, ac became the preferred system for industry. (We'll talk more about motors in a later chapter.)

While we're on the subject of wire sizing, let's run through a simple example of the basic power equation to calculate a required wire size.

Power Technology

Associated with this example are references to the National Electrical Code. **"The Code,"** as it's generally called, is the criteria used by designers and inspectors for the practical safeguarding of persons and property from hazards arising from the use of electricity. The Code is prepared by the National Fire Protection Association (NFPA). By itself, the Code has no force of law. However, just about every jurisdiction in the U.S. has formally adopted it as mandatory.

Now on to our example. Rearrange the power equation as follows:

$$I = \frac{P}{V}$$

For our example we will use a 100-watt (100-W) light bulb supplied by 115 vac.

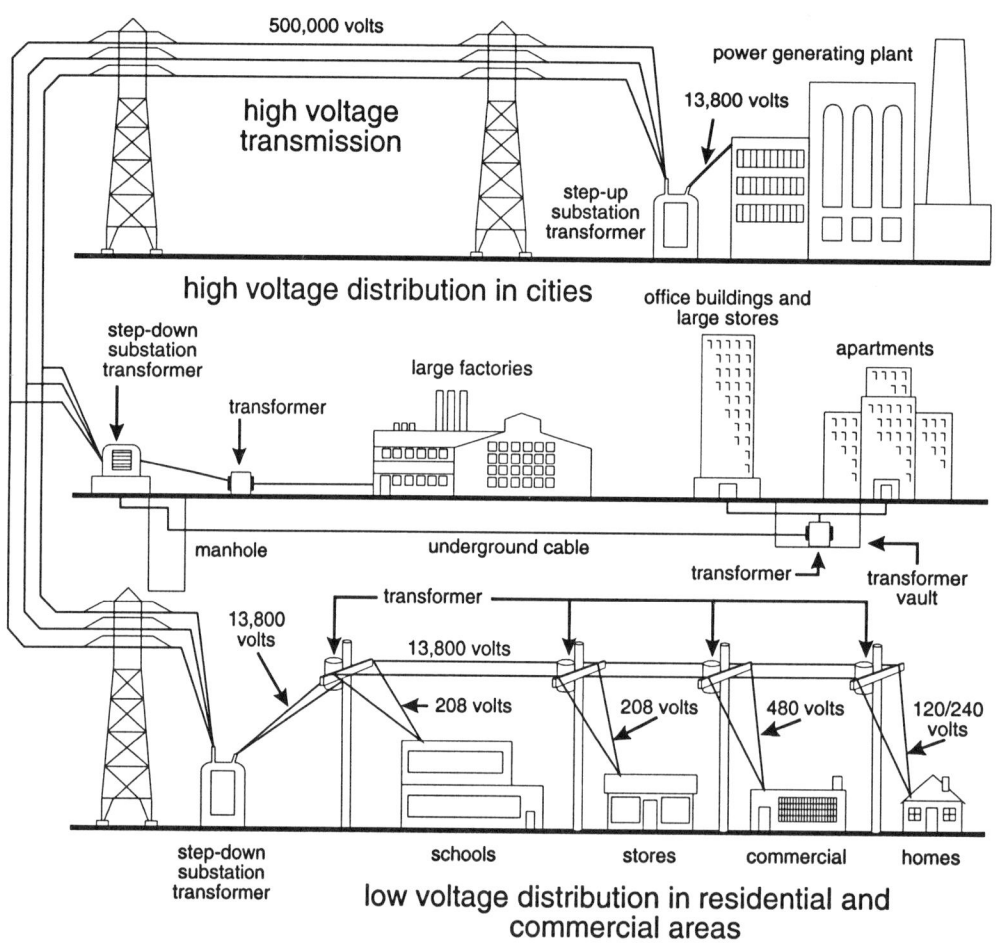

Figure 1-1. Power distribution diagram

To calculate how much current it consumes, divide the power (in watts) by the voltage:

$$I = \frac{P}{V} = \frac{100 \text{ watts}}{115 \text{ volts}} = 0.87 \text{ amps}$$

As elementary as this seems, this is the equation used when calculating the capacity requirements for circuits.

Now let's run through a problem that an electrician might encounter.

Let's say a new plate warmer using a bank of infrared bulbs totalling 1,800 W is to be installed in a cafeteria, and a 15-amp circuit is available. Is this adequate?

$$I = \frac{P}{V} = \frac{1,800 \text{ watts}}{115 \text{ volts}} = 15.7 \text{ amps}$$

You might be tempted to go with the 15-amp circuit, since it is over the circuit availability by just seven-tenths of an amp, but there would be two Code violations with this installation.

First of all, over is over. For even one-tenth of an amp, the circuit must be increased to the next largest size. Second, because this is a continuous load (defined in the Code as three hours or more), the circuit must be derated to 80% of its listed capacity. This means that our 15-amp circuit now is only good for 12 amps (15 amps x 0.80 = 12 amps).

So what size circuit is required? Take our 15.7-amp load and divide it by 0.80 (15.7 amps divided by 0.80 = 19.6 amps). We would need to replace the 15-amp breaker with a 20-amp breaker. If #14-ga wire was connected to the original 15-amp breaker, it would need to be replaced with larger, #12-ga wire.

That's right. The smaller the gauge number, the larger the wire. Let's continue to Chapter 2.

Chapter 2: TRANSFORMERS AND AUTOTRANSFORMERS

TRANSFORMERS

Car transmission systems are very similar in their performance to transformers. Car engines and power station generators both produce power, but not necessarily in the form required for the job.

When a car starts to move, high torque at low rpm is needed at the drive wheels. When cruising, high rpm at low torque is desired. The transmission trades off rpm and torque to match these requirements, but doesn't increase or decrease horsepower from the engine.

$$\text{horsepower} = \frac{\text{torque} \times \text{rpm}}{5252} \text{ (in)} = \frac{\text{torque} \times \text{rpm}}{5252} \text{ (out)}$$

Transformers trade off voltage and current to match requirements, but do not increase or decrease power (watts) from the generator.

watts = volts x amps (in) = volts x amps (out)

Transformers and gear boxes are both efficient at converting power, but neither is perfect. Automatic transmissions require coolers, so the transmission fluid doesn't overheat as it carries away the heat caused by friction. Transformers also require cooling to carry away the heat caused by internal power losses. That's why there are radiator fins on large liquid-cooled transformers, and also why air needs to circulate freely around smaller air-cooled transformers.

A simple transformer consists of two coils: a primary and a secondary, Figure 2-1. The **primary coil** receives energy from the supply source; the **secondary** receives energy from the primary and delivers it to the load. The coils are wrapped around an iron core that increases its efficiency by providing a low-resistance path for the magnetic fields.

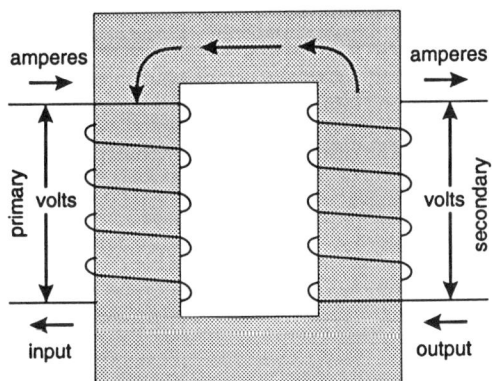

Figure 2-1. A simple transformer

The primary and secondary coils are electrically insulated from one another, so none of the current passing through the primary gets to the secondary. Electricity is generated by the relative motion between a conductor and magnetic field. All the electricity in the secondary coil is generated by the primary's electric field, as its alternating current rises and falls.

Transformers are strictly ac machines. They cannot transform dc. The ignition coil in your car is a transformer. But if transformers can't be used with dc, how does it work? Like this: Dc from the car's battery passes through the primary, except when the distributor points or solid-state switch opens. When the current stops, the established magnetic field rapidly collapses, inducing high voltage in the secondary.

The transformer's **turns ratio** determines the output voltage.

$$\text{turns ratio} = \frac{\text{number of turns in the secondary coil}}{\text{number of turns in the primary coil}}$$

If a transformer has a turns ratio of 2:1, the output voltage is double the input voltage. If it has a turns ratio of 1:2, the output voltage is half the input voltage. The former is referred to as a **step-up transformer**, the latter as a **step-down transformer**.

The step-down transformer in Figure 2-2 has 200 turns in the primary and 10 turns in the secondary, or a turns ratio of 20:1. An input of 2,400 V is reduced to an output of 120 V (2,400 divided by 20 = 120). If the input and output were reversed, it would be a step-up transformer.

In transformers, the current in the secondary is the inverse of the voltage ratio. So, if a transformer has a turns ratio of 2:1, the output current is half of the input current. If it has a turns ratio of 1:2, the output current is double the input current.

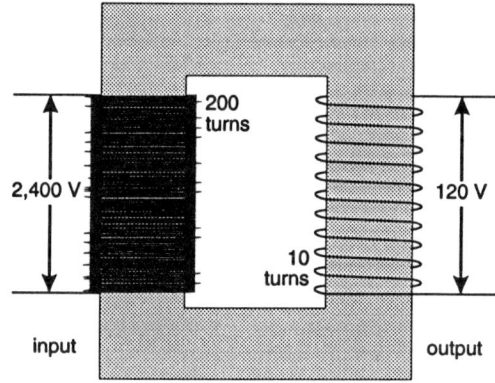

Figure 2-2. A step-down transformer

Let's assume an input of 120 V and 20 amps (20 A) on our 1:20 turns ratio transformer. (Note that the 1 and 20 were turned around to indicate a step-up transformer.) As predicted, the voltage increased 20 times while the current decreased 20 times. Since volts times amps equals power, and since the power entering a transformer should equal the power coming out, this works nicely, Figure 2-3.

120 V x 20 A = 2,400 V x 1 A = 2,400 W

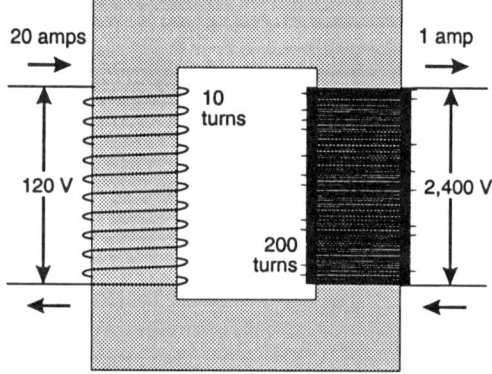

Figure 2-3. A step-up transformer

As stated before, there is no direct electrical connection between the primary and secondary coils of a transformer. When it is desirable to isolate a load electrically from its supply, but not to change the voltage, an **isolation transformer** is used with a 1:1 ratio, Figure 2-4.

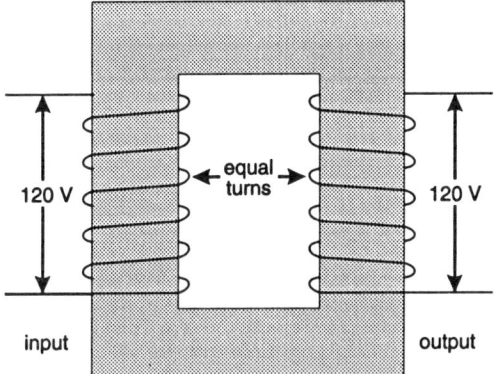

Figure 2-4. An isolation transformer

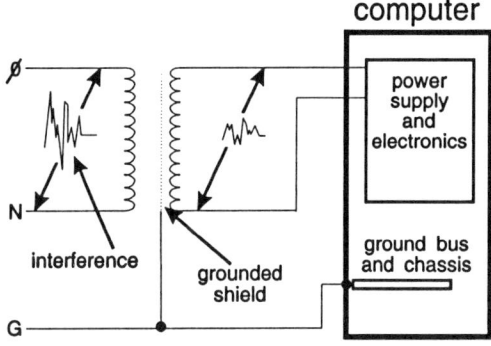

Figure 2-5a. An isolation transformer with a grounded shield

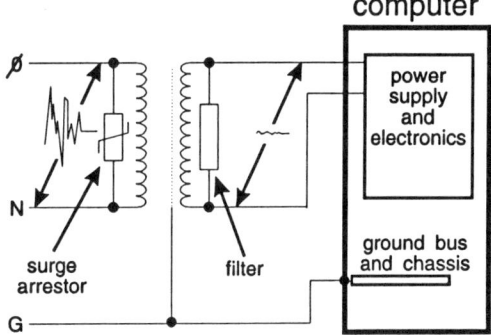

Figure 2-5b. An isolation transformer with a surge arrestor and filter

Some power systems have voltage spikes. Still others create "noise" that interferes with the proper operation of electronic equipment. Isolation transformers electrically segregate the more delicate power-consuming equipment, by placing a grounded shield between the primary and secondary, Figure 2-5a. This shorts out to ground the radio-frequency interference before it reaches the load. ("Shorts out" means that a very low resistance or easy path to ground is provided.) If it were allowed to reach the load, this type of interference would be harmful to accurate data communication between equipment such as computers.

Sometimes surge arrestors and filters are added to isolation transformers, to eliminate these power line interrupters, Figure 2-5b. A surge suppressor is a device that shorts out at a predetermined voltage level, then restores itself when voltage returns to normal. That way, harmful voltages never reach the equipment. Filters can be designed to short out harmful frequencies, while letting through the normal, 60-cycle power.

Wide use of isolation transformers helps ensure personal safety from electrical shock, by isolating the incoming primary winding power lines to ground. One ground will not cause a problem on the secondary of an isolation transformer.

SIZE AND COOLING

The first step in transformer selection is determining the power required in VA (volt-amps) or kVA (kilovolt-amps). If the power was given in watts, or real power, it would be only one component of the load. The other component is reactive or magnetizing power. Volt-amps, or apparent power, is the combined measurement of the two, and it defines the total load of the transformer. (The topics of real and reactive power will be covered in more detail in Chapter 6.)

Another consideration is how the transformer is going to be cooled. Heat is what limits a transformer's capacity. The more efficient the heat removal, the more power a transformer can deliver. Air-cooled types are the least expensive and work fine for smaller units. Larger units require liquid cooling to remove heat from the windings.

Mineral oil is an efficient transformer coolant. However, because mineral oil is flammable, it is usually limited to outdoor installations. Fire-resistant coolants such as Askarel were developed for indoor use, but are now a liability because of their high PCB content.

AUTOTRANSFORMERS

Until now, all the transformers discussed have had separate primary and secondary windings. **Autotransformers** combine both in the same winding. Their main advantage is that they provide an inexpensive and efficient way to change voltage. One drawback is that they provide no electrical isolation between primary and secondary windings.

Autotransformers are commonly used to match equipment voltage requirements with the available supply. For example, say that a 240-V motor was purchased, but only 208-V power is available. Instead of purchasing a two-winding transformer to carry the entire 240-V load, an autotransformer can save a considerable amount of money because it only needs to be sized for the additional 32 V, Figure 2-6.

Figure 2-7 shows a common way to use two autotransformers to "buck" or "boost" the voltage in a three-phase load. (Buck and boost are terms associated with autotransformers; **boost** means to slightly increase voltage, **buck** means to slightly decrease voltage, to match an application, Figure 2-8.)

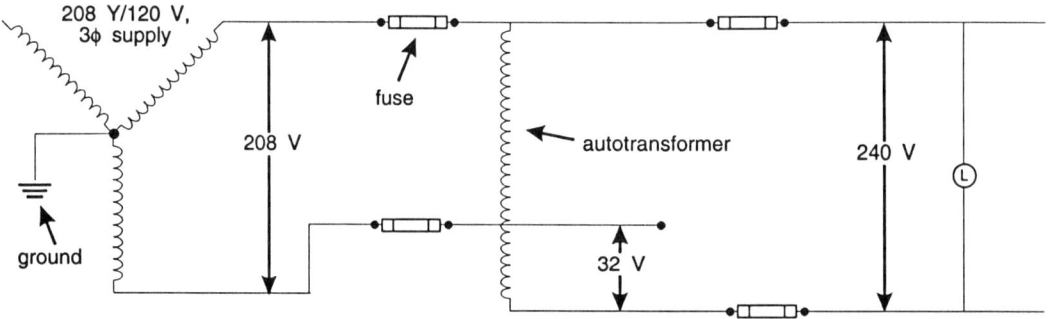

Figure 2-6. Autotransformer

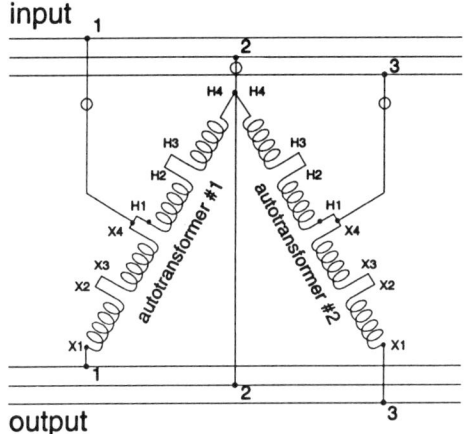

Figure 2-7. Two autotransformers used to buck or boost voltage

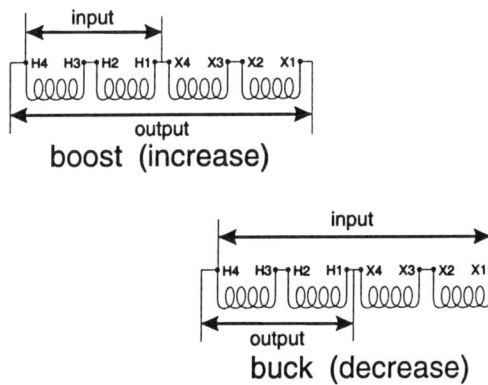

Figure 2-8. Buck and Boost

Chapter 3
SINGLE- AND THREE-PHASE POWER

Sooner or later, when discussing electric power, the terms **single phase** and **three phase** will come up. The description of single-phase power is fairly straightforward; three-phase power is slightly more complicated.

To demonstrate the configuration of single-phase power, imagine yourself with arms outstretched from your sides, Figure 3-1.

Note: *For the proper electrical terminology, substitute the word* **line** *for finger tips, and* **neutral** *for chin.*

The distance between chin and finger tips would be around 3 ft; the distance from finger tips to finger tips would be around 6 ft. This is similar to the configuration of a single-phase transformer (like the one that supplies your house): 120 V from finger tips to chin, 240 V from finger tips to finger tips.

To demonstrate the configuration of three-phase power, imagine yourself raising your hands until your arms are 30 degrees above horizontal, forming a 120-degree angle, Figure 3-2.

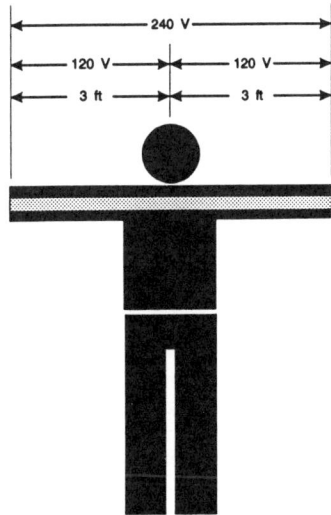

Figure 3-1. Demonstration of single-phase power

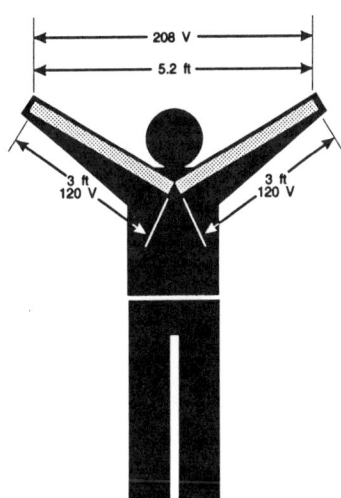

Figure 3-2. Demonstration of three-phase power

Power Technology

As the hands go up, the distance from finger tips to finger tips decreases to about 5.2 ft, while the distance from finger tips to chin remains 3 ft. This is like the configuration of a three-phase transformer (generally used to supply offices and factories): 120 V from finger tips to chin, 208 V from finger tips to finger tips. Working out the trigonometry of a 120-, 30-, 30-degree triangle, the longer side is 1.73 times longer than the shorter sides; 1.73 — square root of three — is used frequently when calculating three-phase problems.

SINGLE-PHASE POWER

Most homes have 120/240-V service, as schematically represented in Figure 3-3. The transformer in the street supplies 240 V between **L1** and **L2** (**line one** and **line two**). You can measure 120 V between L1 or L2 and the neutral. The neutral is tapped off of the center of the transformer's secondary. The center-tap neutral is grounded at the transformer and in the service entrance panel.

Notice that the incoming house power goes through the electric meter service entrance panel. Removing the meter disconnects the power. A main circuit breaker protects the panel from potential electrical damage, and disconnects power when work needs to be done inside the panel.

A **bus bar** is a flat, rectangular, copper or aluminum bar that carries current in electrical equipment such as panels and switchboards. It can be punched, or drilled and tapped, to accept various electrical connections.

Each circuit breaker connects to one of two bus bars in the panel. Adjacent breakers are connected to alternate bus bars; 120-V circuits use one circuit breaker and the neutral. For 240-V circuits, two breakers

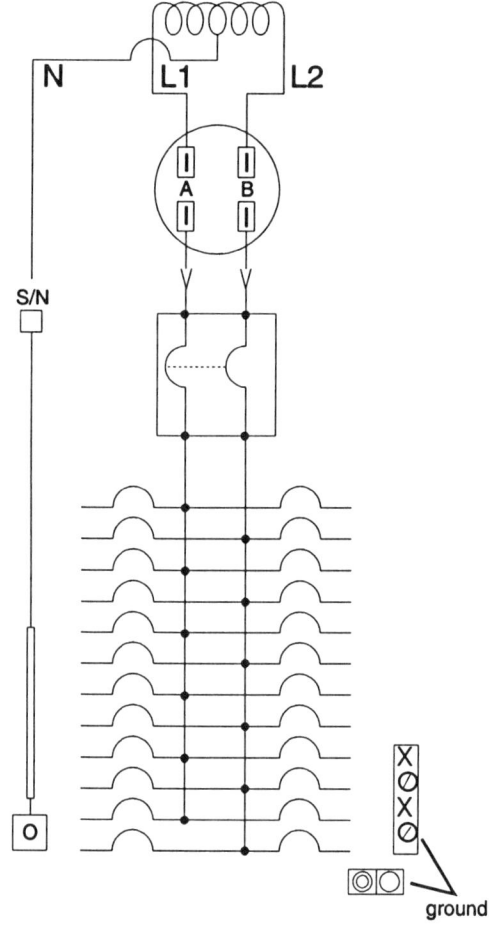

Figure 3-3. 120/240 volt service transformer

are **ganged** together to form a single unit. (The neutral is not used when 240-V power is required.) Each of the two ganged breakers makes contact with a different bus bar so, if one of these two-pole breakers is turned off or trips, both poles operate as a unit so all power is removed from the circuit.

There also are a neutral bus and a ground bus. These provide attachment points when the neutral and ground wires are connected.

When the front cover is removed from a power panel, the bus bar and circuit breaker arrangement becomes more comprehensible. However, uncovered panels make me nervous. People who take a nonchalant attitude when working on parts that have the potential of being energized, are putting themselves and others in danger. Make sure you read the tagout-lockout and safety procedures covered in Chapter 4.

Figures 3-4a and 3-4b show how single-pole and double-pole breakers are connected to a panel to feed 120- and 240-V circuits.

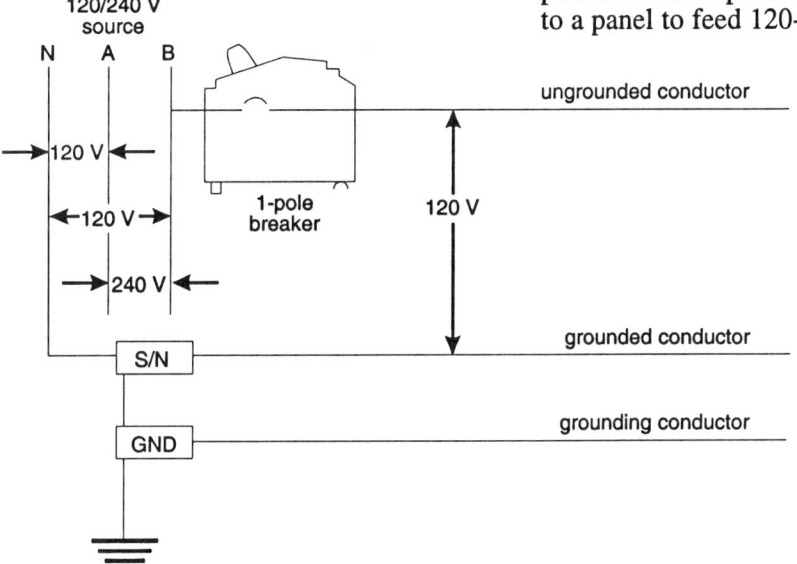

Figure 3-4a. Single-pole breaker

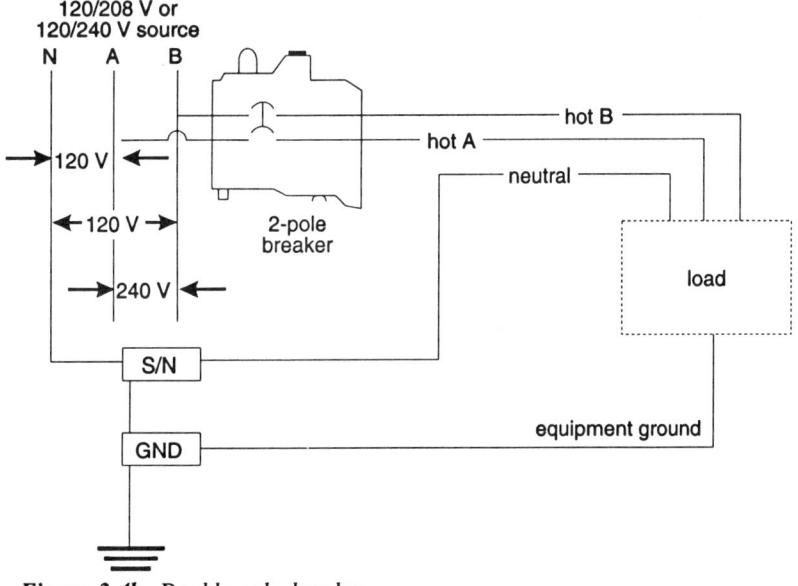

Figure 3-4b. Double-pole breaker

Although there is no mandatory color code for energized conductors, the wires connected to breakers for 120/240-V circuits usually have black- or red-colored insulation. Neutral wires must have white insulation, and ground wires must have green insulation.

When loads are turned on or off, like a light, it is always the "hot" (black or red) conductor that is switched. Neutral and ground are never switched.

Notice the terminology used for neutral and ground wires. The neutral is called the grounded conductor, and the ground is called the grounding conductor. In service equipment panels, where the electric service enters the building, the neutral and grounded conductors are bonded or connected together. In subpanels — power panels that are fed from the main panel — the neutral bus bar is isolated from the enclosure, and no connection is made between neutral and ground.

Minor power users, like TVs and toasters, use 120 V. One blade of the plug is the hot leg and is connected to a single-pole breaker. The other blade is connected to the neutral. The wider of the two blades is the neutral.

Appliances with large power appetites, like air conditioners and stoves, are connected to both hot ungrounded conductors for 240-V operation. (Remember, for a given amount of power, when the voltage is doubled, the required current is halved.) Both ungrounded conductors are connected to a two-pole breaker. Each pole of the two-pole breaker connects to a different bus bar in the panel. The handles of the two-pole breaker are tied together so that they can operate as one unit.

As a point of interest, most European countries use 240-V power in their homes for common appliances. It is their philosophy to minimize conductor material by using the higher voltage.

THREE-PHASE POWER

Three-phase power is preferred by power companies and larger power users. Why? For starters, the rating of three-phase motors and generators is almost 50% greater than when operating single phase. Associated equipment also is smaller and less complicated. Poly-phase motors especially are simpler and less expensive than single-phase units.

Another benefit is that three-phase power costs less to transport. Compared with a single-phase system delivering an equal amount of power, three-phase system conductors weigh 75% less.

Single-phase power pulsates. It can be compared to a single-cylinder engine, where the flywheel returns energy to the cylinder during the compression part of the cycle. Three-phase power can be compared to a multi-cylinder gasoline engine, where the power delivered to the flywheel is steady, as one cylinder fires when the others are compressing.

Figure 3-5 shows a schematic representation of a three-phase power panel. The breakers can be connected to one of three energized bus bars, instead of the two found in the single-phase panel. Like the single-phase panel, adjacent circuit breakers are connected to alternate bus bars. There also are neutral and ground bus bars in three-phase panels.

The conductors of different voltage systems must be identified with different col-

Single- and Three-Phase Power

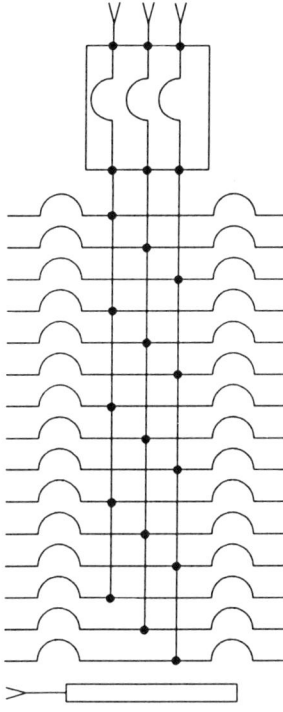

Figure 3-5. Schematic representation of a three-phase power panel

ors. There is no mandated standard, but the following is common:

- For 208/120-V systems, the wires connected to the circuit breakers have black, red, and blue insulation.

- For 480/277-V systems, the wires connected to the circuit breakers have brown, orange, and yellow insulation. Brown, Orange, Yellow is often referred to as the BOY system.

As in the single-phase system, the neutral and ground conductors are not switched; they have an unbroken path back to the panel, regardless if the breakers and switches for the circuit are on or off. The neutral conductors in 480/277-V systems sometimes are gray, to distinguish them from 208/120-V systems. Grounding conductors are always green.

Let's review how we got these 480/277- and 208/120-V systems. Remember, the magic number for three-phase systems is 1.73 – the square root of 3.

For 480/277-V systems, 3φ (a.k.a., three phase) is a desirable distribution voltage. (The Greek letter *Phi* (φ) is the symbol used for phase.) It is economical to install, electricians are familiar with it, and it is the standard for motors and heating equipment. Lighting systems also can operate directly off 480/277 voltage, eliminating the need for additional transformer capacity. To save panel space and money for breakers, lighting is connected as a single-phase load, from a phase to the neutral. The voltage from phase to neutral is 480 divided by 1.73; 480 divided by 1.73 equals 277 V.

As useful as 480/277 voltage is, it can't satisfy the tremendous amount of equipment that requires 120-V power. To fulfill this, a transformer is required. The transformer's primary coil is fed from the 3φ, 480-V riser, and the secondary coil feeds a 208/120-V, 3φ receptacle power panel. The single-phase, 120-V outlet power is connected to single-pole breakers and neutral.

Figure 3-6 shows the distribution system in a typical office building. It shows two switchboards feeding two 480-V, 1,600-A risers. On each floor, each riser feeds a 480/277-V panel for lighting and other heavy, three-phase loads; and a 480-208/120-V step-down transformer feeds a second panel for 120-V outlets and other light loads.

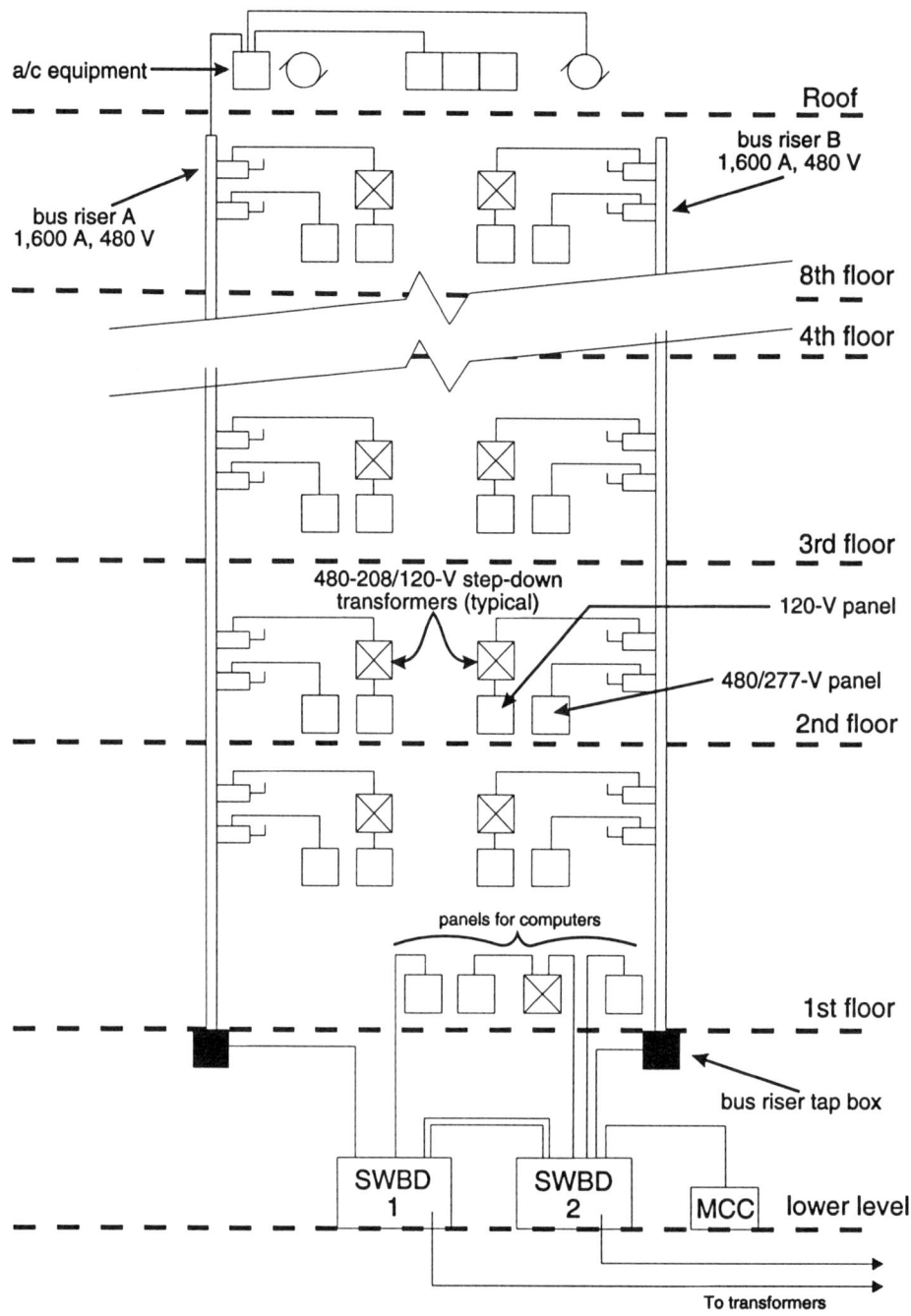

The electric power distribution within the building is divided in half, with service Switchboard 1 feeding bus riser A, and Service Switchboard 2 feeding bus riser B. This arrangement is typical for office buildings.

Figure 3-6. Distribution system for typical office building

Pictured in Figure 3-7a is a **wye configuration**. If no neutral is used, it is called a **three-wire system**. If a neutral is connected to the common point, it is called a **four-wire system**. If a ground is used, it becomes a **five-wire system**.

Figure 3-7b shows another common three-phase connection: the **delta configuration**.

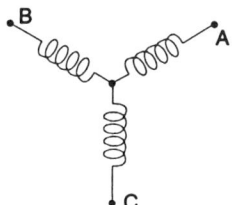

Figure 3-7a. Wye configuration

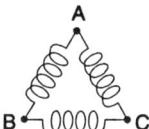

Figure 3-7b. Delta configuration

Most transformers have a delta primary, with a wye or delta secondary.

The wye secondary is popular when there is a high percentage of single-phase loads such as 120-V equipment and lighting.

If there are mostly line-to-line loads, with a small percentage of line-to-neutral requirements, the **center-tapped delta** is used. As shown in Figure 3-7c, each winding supplies 240 V, but with one winding center tapped, 120/240-V single-phase power can be supplied. Also, 208 V is present between the center tap and the high leg. The high leg is always the "B" phase in the power panel, and the insulation is always orange for identification.

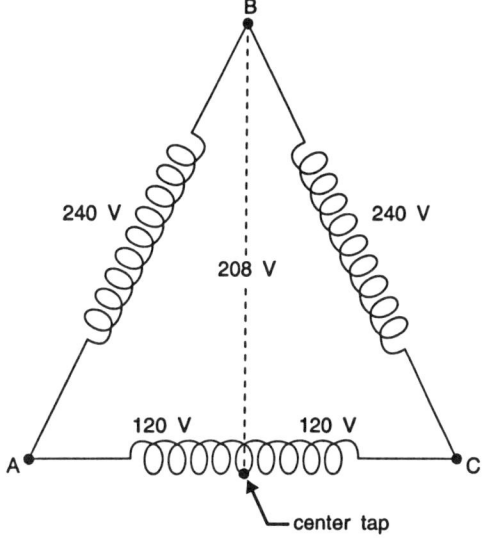

Figure 3-7c. Transformer with one center-tapped winding

The open delta is popular because it saves the expense of a third transformer, while being capable of an easy upgrade if needed, Figure 3-8.

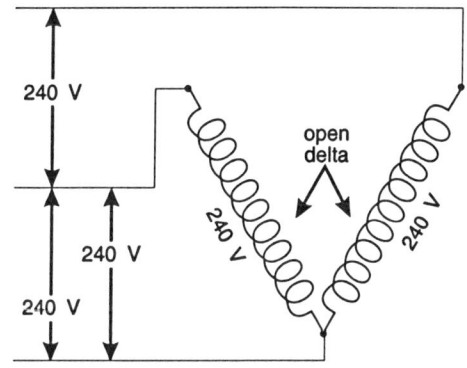

Figure 3-8. V or open-delta connection of transformers

Say, for example, that a light industrial park is being developed on speculation, and the power company has no firm projection of future demand. Instead of tieing up capital by installing transformer capacity that might never be used, you can install an open delta system initially, and add a third transformer later if it's required. The project immediately becomes more feasible, and the bank sees greater potential to get its investment back.

Another reason for this arrangement is its versatility if one of the transformers fails. By taking out the defective unit and wiring the other two in an open delta configuration, 58% of the original capacity is available until repairs can be made.

HAZARD PREVENTION AND SAFETY

When working with electrical systems, hazard prevention and safety come first. This topic can be broken down into two main divisions: protecting people from injury (personal protection) and protecting equipment from damage or destruction. Because it is nearly impossible to talk about one without talking about the other, this chapter will flip back and forth between personal protection and equipment protection.

PLUGS AND OUTLETS

Because they are so common, we tend to take plugs and outlets for granted. However, if you plugged an appliance rated for 120 V into a 240-V outlet, the result would be devastating to the appliance and hazardous for you. Fortunately, all plugs and outlets have specific voltage and current ratings. Table 4-1 gives you a sample of the various types of plug and outlet configurations available. It is only intended to show a few configurations, with their voltage and amp ratings.

Note: A NEMA number is associated with each configuration. The National Electrical Manufacturers Association, along with the American National Standards Institute (ANSI), set standards for safety and performance. So, if you need a 6-15R or L10-20P, the configuration will be the same regardless of its manufacturer.

> **FOR YOUR INFORMATION —**
> When specifying a configuration:
> - The prefix **"L"** means **twist lock**.
> - The suffix **"R"** means **receptacle**.
> - The suffix **"P"** means **plug**.

An **isolated ground receptacle outlet** is a specialized outlet used frequently in computer centers. It is used because there usually is some stray current flowing in ground wires and in the building steel. Occasionally, the current is high enough to cause problems in sensitive equipment, where multiple ground paths or ground loops exist. Isolated ground outlets prevent these ground loops by providing an isolated, dedicated ground path back to a main distribution panel.

With **three-prong receptacles**, the green-colored ground screw and round ground prong are connected. The mounting taps on conventional receptacles also are connected to the green-colored ground screw. When screwed into a metal box, the unit eventually comes in contact with other grounded metal parts. Under these circumstances, ground loops result.

Power Technology

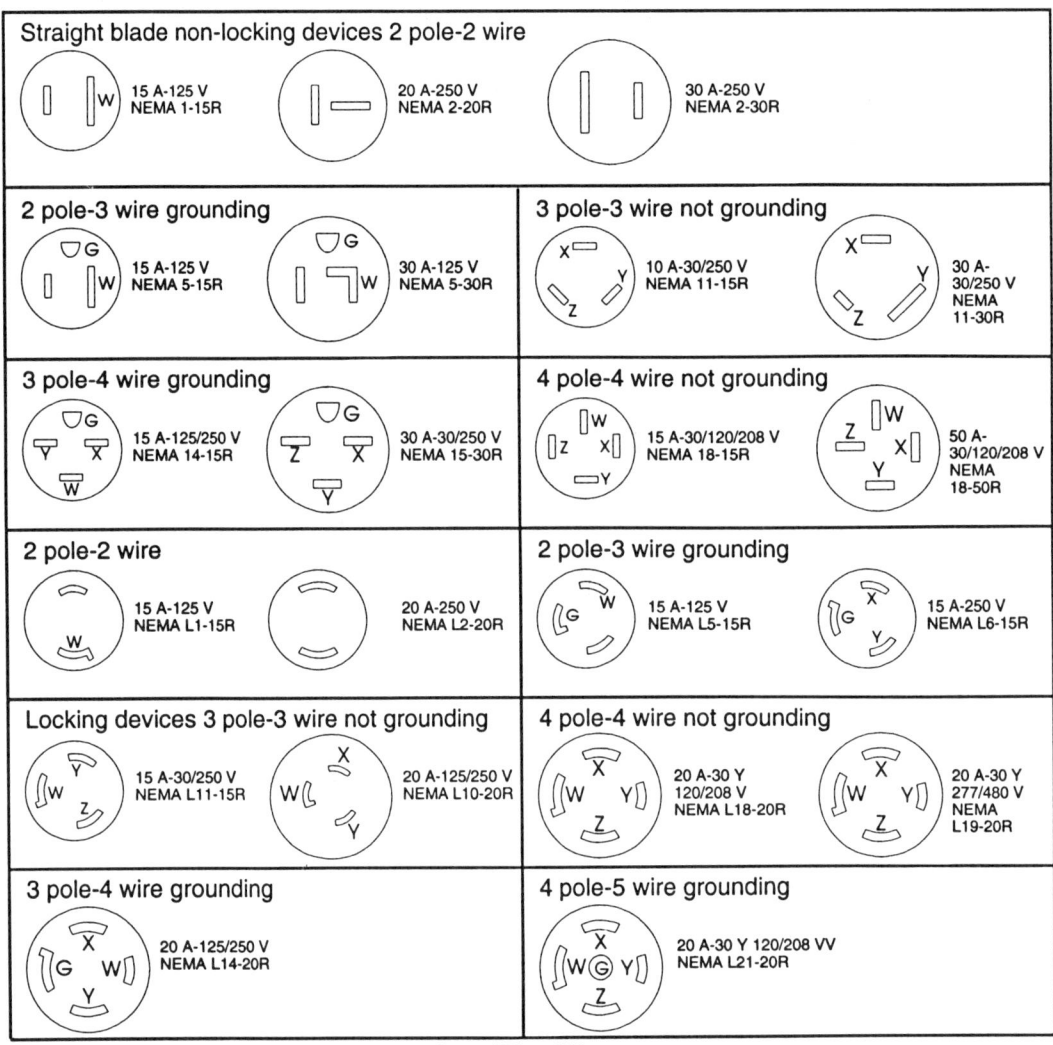

Table 4-1. Examples of various plug and outlet configurations

The mounting tabs and ground prong on isolated ground outlets are not connected. When an insulated ground wire is used, the equipment ground is isolated all the way back to the power panel. In fact, the ground can go through the panel without making any connection and can continue to the next upstream panel. This may even continue until the panel nearest the power source is reached.

Now let's discuss exposed metal surfaces on the junction box and the rest of the system. As shown in Figure 4-1, a second ground connected to the first upstream panel is used to comply with Electrical Code

Hazard Prevention and Safety

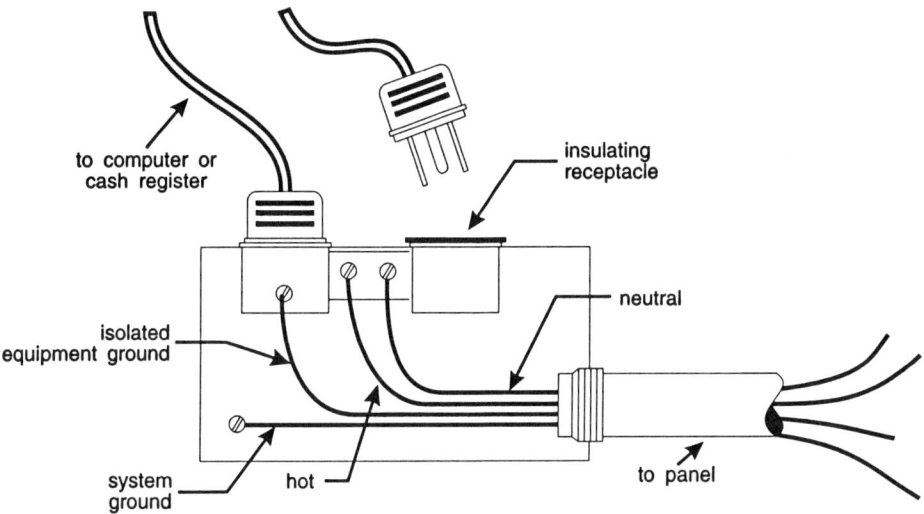

Figure 4-1. Isolated ground receptacle with two ground conductors

safety requirements. There may be ground loops in this ground circuit, but they will remain isolated from the critical equipment. The illustration shows how an isolated ground receptacle is wired using two ground conductors.

ONE-LINE DIAGRAMS

One-line diagrams show a large amount of information in a clear, compact form. They are called one-line diagrams because a single line is used to represent all the phase conductors, neutrals, and grounds.

Having the information concisely presented helps operators plan and understand which procedure to follow when switching feeders or securing panel boards; or helps them find out why a particular building is without power.

Figure 4-2 shows a typical one-line diagram. In a small amount of space, it shows a service entrance with a 1,500-kVA (kilo-volt-ampere) transformer, with a 13.2-kV delta primary and 480/277-V grounded wye secondary, protected with a type EJ-1 fuse. The bus is protected with 2,500-A circuit breaker that also can be tripped with a ground fault protector.

An additional note about one-line diagrams: All conductors are shown with sizes and lengths. Usually one-lines stop at power panels and motors.

Engineers use one-lines for coordination studies and fault current evaluations.

Coordination studies: Every circuit breaker and fuse has its own unique characteristic. Coordination studies orchestrate protection devices, so that the circuit breaker or fuse closest to the fault opens first.

Fault currents: A short circuit occurs when a live conductor comes in contact with another conductor or ground. A fault is a short circuit that occurs on high-energy

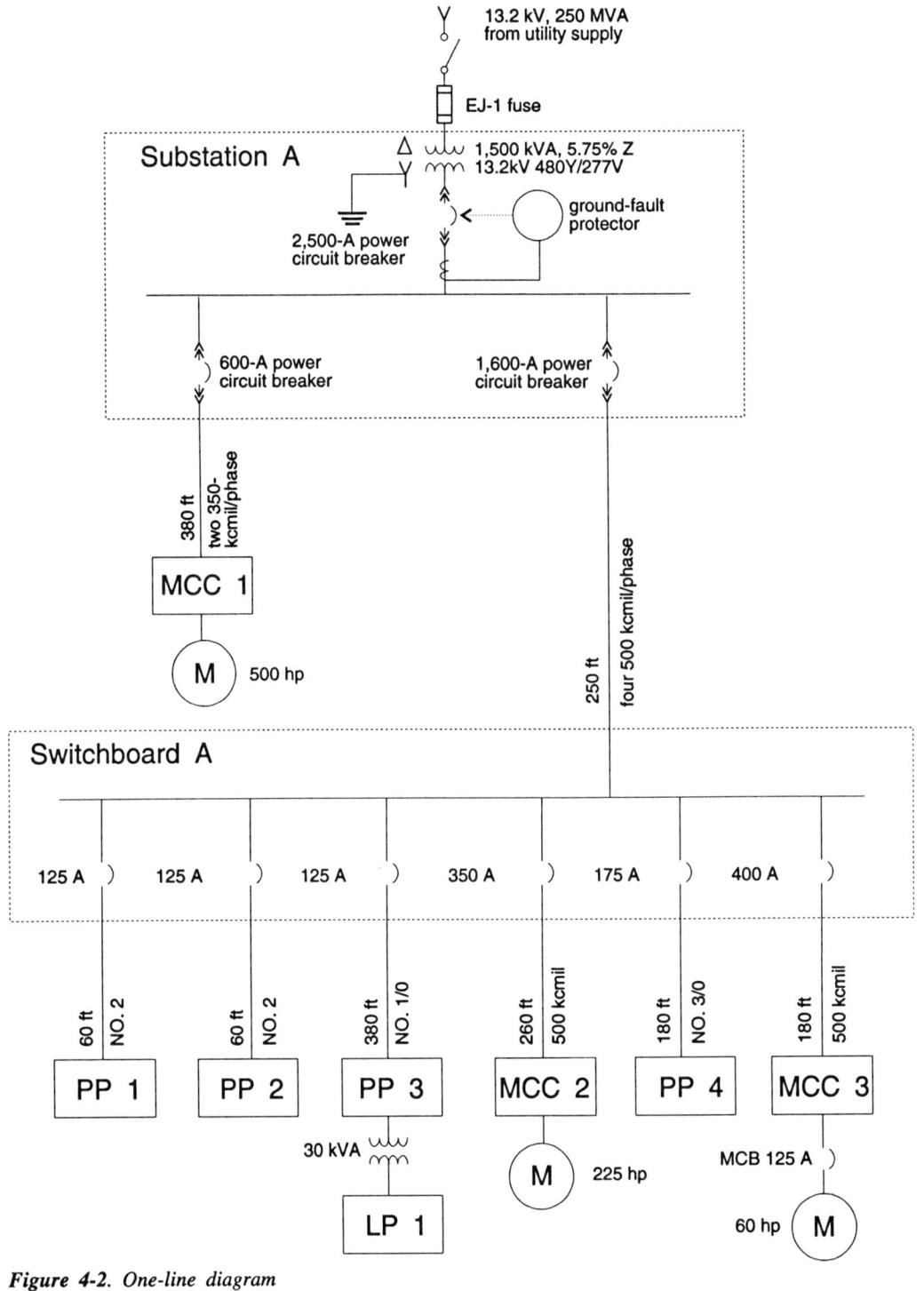

Figure 4-2. One-line diagram

circuits. A fault can be explained by drawing an analogy to a hydroelectric dam. During normal operation, a regulated amount of water is allowed to run through the plant to turn the turbines. If the dam breaks, all of the energy within the normally placid stored water is turned loose, causing havoc downstream.

Basically the same thing happens when a fault occurs. The entire capacity of the power system is available to feed all the current it can take. If the source of power is not cut off instantly, the damage can be disastrous. Protective devices like fuses and circuit breakers must not only open at their rated amperage, but also be capable of interrupting large fault currents.

Think of it as the difference between standard- and heavy-duty brakes. Both are capable of stopping a truck on a level road, but it's a different story when controlling a fully loaded truck on a long, steep downgrade. Standard-duty brakes may fail with disastrous consequences. Likewise, when a fuse or circuit breaker can't handle the fault current, they fail, potentially causing an explosion.

ELECTRICAL SHOCK

Current is the killing factor in electrical shock. Voltage and electrical resistance determine how much current flows through the body. Figure 4-3 depicts how Ohm's Law determines how much current passes through the body.

An ohm is a unit of electrical resistance. One volt applied across a resistance of one ohm, causes a current of one ampere. The symbol for ohm is the Greek letter *Omega* (Ω).

Ohm's Law explains the relationship between voltage, current (amps), and resistance (ohms). It states that the amount of current through a conductor is directly proportional to voltage applied and inversely proportional to the resistance of the conductor or current.

$$\text{amps} = \frac{\text{volts}}{\text{ohms}}$$

If 110 V is placed across a 500-ohm resistor, the resulting current would be 0.22 A, or 220 mA. (The m stand for milli, or one-one thousandth (1/1,000). 220 mA is pronounced, 220 milliamps.)

$$\frac{110 \text{ V}}{500 \text{ ohms}} = 0.22 \text{ A, or } 220 \text{ mA}$$

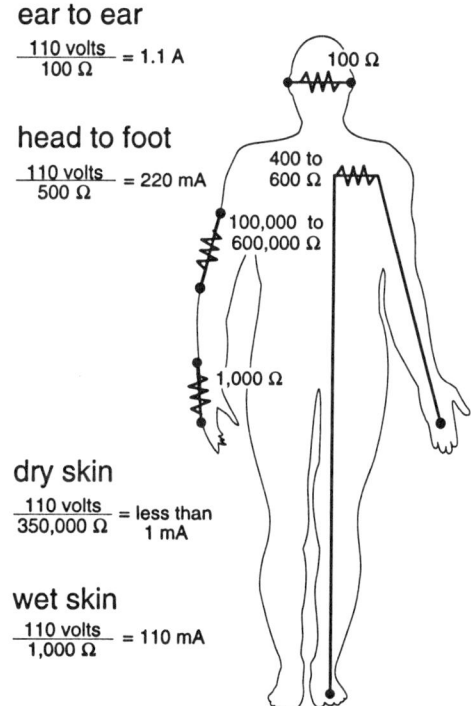

Figure 4-3. How Ohm's Law determines how much current passes through the body

As shown in Figure 4-3, human bodies have resistance. When voltage is applied between two locations on a body, current flows.

Most people are killed by 110-V power, probably because we all tend to take it for granted. Keep in mind, however, that *it is lethal if only one-tenth of the current required to operate a 10-W light bulb passes through your chest cavity.*

SHOCKING NEWS!

Here's what you'd feel if you picked up an appliance radiating:

- Less than 1 mA — No sensation; nothing felt.

- 1 to 8 mA — Sensation of shock, but not painful. You can let go at will; muscle control is not yet lost.

- 8 to 15 mA — Painful shock, but you can still let go at will.

- 15 to 20 mA — Painful shock. Can't let go because control of adjacent muscles is lost.

- 20 to 50 mA — Severe muscular contractions. Breathing is difficult. Suffocation possible.

- 100 to 200 mA — Ventricular fibrillation, a heart condition that can result in death within minutes.

- Over 200 mA — Severe burns and muscle contractions. Chest muscles clamp heart and stop it during duration of shock.

LOCKOUT-TAGOUT PROCEDURES

One of the most-effective ways to prevent electrical shocks is with an aggressive, effective lockout-tagout program. And it's no longer just a matter of it being a nice thing to have. **It's the law.**

The concept really is simple: Whenever a piece of equipment is being worked on, and if it's still connected to an energy source, that energy source must be locked out to prevent accidental activation. This not only includes electricity, but also other energy sources, like steam.

The person who locked out the source should have the only key. If more than one person is working on the equipment, each person should have a lock in place. Some devices on the market accommodate multiple locks.

In addition, the energy source must be tagged with the following information:

- Who is working on the equipment;
- What is being done;
- Why it's being done;
- When.

PERSONAL GROUND FAULT PROTECTION

Ordinary circuit breakers don't protect people against shock. Fortunately, **Ground Fault Circuit Interrupters (GFCI)** have been developed. Figure 4-4 shows how a ground fault circuit interrupter works.

The current balance in the hot conductor and neutral is continuously monitored. Under normal circumstances they are equal and cancel each other out. However, if the current in the neutral is less than the current in the hot conductor, a ground fault exists; some of the current is diverted to ground, such as when someone or something is getting a jolt. If this difference is greater than 6 mA, the circuit opens.

There is a growing list of locations where GFCIs are now required, including bathrooms, construction sites, garages, kitchens, marinas, and pools.

A ground fault can be simulated by pushing the test button. If the breaker doesn't trip when tested, replace it.

GROUND FAULT PROTECTION FOR EQUIPMENT

Ground fault systems also protect power systems and equipment from potentially catastrophic damage caused by the enormous amount of current ground faults create. Entire switchboards, along with the associated cable, have been wiped out from ground faults. Arcing from a 480/277-V system is particularly destructive. For this reason, protection is required on 480/277-V services rated at 1,000 A or more.

The ground fault equipment is installed at the service entrance where power first enters the building. In this case, the setting can be as high as 1,200 A, with a maximum time delay of 1 sec. When a ground fault occurs, the whole plant goes down.

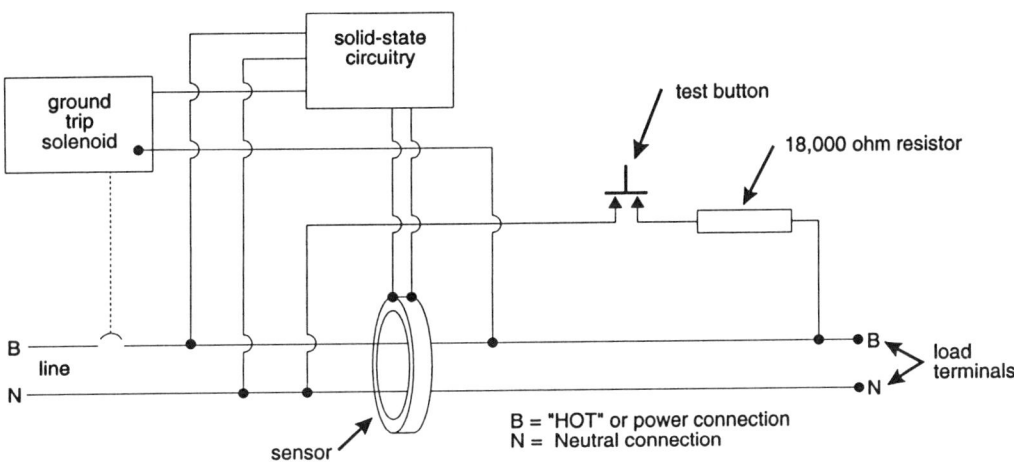

Figure 4-4. Ground fault circuit interrupters

Chapter 5

DEMAND AND THE ELECTRIC BILL

When discussing power technology and electricity, the electric bill should not be overlooked. After all, lowering the bill is one of the consumer's primary concerns. You should be able to look at a utility bill, and determine where those wasteful energy dollars are being spent and how they might be saved.

Before we examine the electric bill, however, let's take a look at the demand charge. This major portion of the bill isn't always fully understood.

DEMAND CHARGES

Why are demand charges added to energy charges? Take a look at Figure 5-1. In both cases 1,200 **kWh (kilowatt-hours)** are used during the day, but in the second case most of the power is used in just one hour. The power company is expected to have enough capacity to meet this demand whenever it's called for — even if most other people are using power at that time (i.e., heavy use of home air conditioning after 5 p.m.). This is often called **peak** power consumption.

The problem is, a lot of expensive power-generating equipment sits around idle and earns no income when the demand is low — say, after 10 p.m. — also called **off-peak** power consumption. Therefore, state utility commissions allow power companies to levy a demand charge along with an energy charge. By levying a demand charge on peak power consumption, the utility helps to balance its income against the off-peak hours.

Demand measures how quickly power is used over a period of time. It is based on an average load over an interval of time,

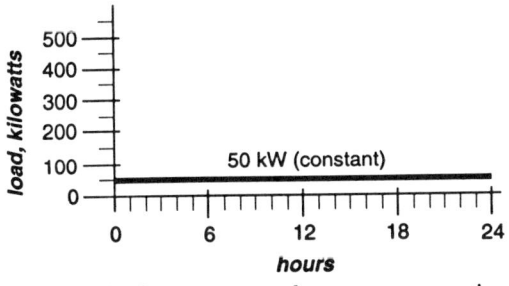
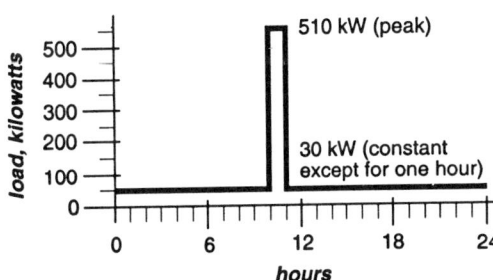

Figure 5-1. Constant vs peak power consumption (1,200 kW total)

usually 15 or 30 min. Figure 5-2 is based on a 15-min demand interval. It shows a 100-kW load, 400-kW load, and 1,000-kW load, which are on for a period of 5 min each. The average load here is 500 kW and that's what the demand charge is based on.

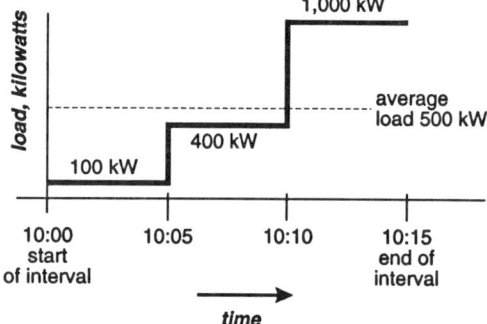

Figure 5-2. Power consumption based on a 15-min demand interval

Some companies take the highest demand for the month; some average out the highest three, but regardless of the method, the average becomes the basis of the demand charge. For example, if the on-peak demand is $10/kW, then 500 kW x $10/kW, or $5,000 is added to the bill.

Demand use can get more complicated. If there is high peak seasonal usage, such as heavy air conditioning use during summer, a higher peak rate is charged, to encourage less use or the installation of more-efficient air conditioning equipment.

There is a common misconception about demand charges related to starting up large motor loads, such as chillers and fans. While it's true that a motor coming up to speed uses six times its normal current draw, it's not true that this starting current has any great effect on demand. For all practical purposes, this starting current is "invisible" to the demand meter. What the meter does detect, is the amount of power consumed by energized equipment over the 15- or 30-min demand period.

In Figure 5-3, two loads are operated 10 min each over a 30-min demand period. It didn't matter if they were operated separately or together. Since they both ran during the same demand period, the demand charges were the same.

In an effort to reduce demand charges, some operators estimate how much air conditioning is required for the peak load in the afternoon, then mistakenly turn on enough equipment during the morning to meet this future demand. This practice does nothing to reduce peak-hour demand, but it definitely adds kWh to the electric bill.

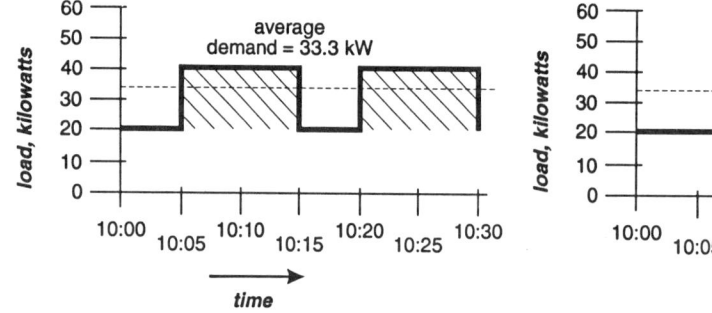
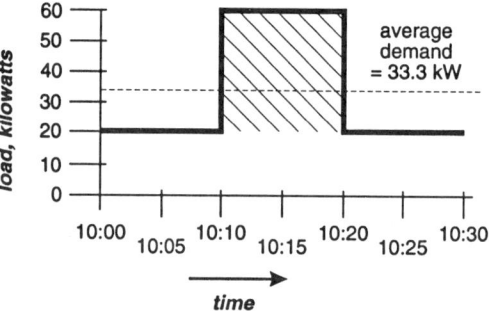

Figure 5-3. Two loads operated over a 30-min demand period

CALCULATING DEMAND

Utility companies can use many methods to calculate demand. Instead of the highest monthly peak, some utilities average out the highest three.

Some utilities have a ratcheting demand clause; a nasty way to keep customers in line. There are various ways that it is calculated, but what it boils down to is that the customer is stuck paying the utility's highest monthly demand rate for a year. An inadvertent high month can sap a company's incentive to get back on the right track.

To calculate a sample demand charge, let's say that a monthly bill from a large office complex came to $28,066 for 335,125-kWh worth of electricity. The total amount due divided by the total kWh yields:

$28,066 ÷ 335,125 kWh = $0.084 per kWh

Using 8.4¢ to calculate demand, however, will lead to erroneous conclusions; it doesn't reflect the time-of-day demand charge. Now let's say that the boss wants to know how much it would cost, per month, to operate an additional 50 kW for eight hours during the day (22 additional working days per month). You figure out this amount with the following calculation:

50 kW x 22 days/month x 8 hr/day x $0.084/kWh = $739.20/month

But without the demand charge figured in, this amount is incorrect. Let's see how much it really costs.

Energy Charge:

50 kW x 22 days/month x 8 hr/day x $0.06744/kWh = $593.47/month

Demand Charge:

50 kW x $9.43/kW/month = $471.50/month

Total: $1,064.97

When the first bill comes in $325 more than expected, the additional demand charge is the reason why. The actual on-peak rate for this load is closer to 12¢ per kWh.

What if that same load was operated off peak?

50 kW x 22 days/month x 8 hr/day x 0.05206 = $458.13/month

You can save the company more than $600 a month by recommending that the boss operate this load at night!

The moral is, when evaluating the economies for either additional loads or purposed energy conservation, calculate all the elements of the rate structure.

PARTS AND PIECES OF THE ELECTRIC BILL

Now for the bill itself. The following items are commonly shown on a typical electric utility statement from a large office complex:

- Energy consumption in kWh;
- Demand in kilowatts (kW);
- Power factor in kilovolt-amperes (kVA) or kilovolt-amperes reactive (kVAR);
- Service charge;
- Fuel adjustment charge.

The following criteria determines how much the consumer pays for the items listed above:

- Primary or secondary service;
- Customer ownership of transformers and equipment;

- Time-of-day rates;
- Seasonal rates;
- Fuel adjustment charge;
- Curtailable service provisions.

Here we have reproduced the items as they might appear on a utility bill:

Rate Classification — General Service at Transmission Voltage

Energy Charges			
On-peak	164,500 kWh @	0.067440	11,093.88
Off-peak	170,625 kWh @	0.052060	8,882.74
			19,976.62
Demand Charges			
On-peak demand	947.10 kW @	9.43	8,931.15
Extra-kVA demand	137.30 kVA @	0.61	83.75
			9,014.90
Service Charge			436.00
			29,427.52
Energy Adjustment Charge 0.004060 per kWh			-1,360.61
			28,066.91

For this utility, general service at transmission voltage means that the service is tapped directly from one of the utility company's transmission lines, and that the customer owns the switch gear and transformers. In return, the utility sells the customer electricity at the cheapest rate it offers. Budgeted moneys need to be provided (by the consumer) for equipment maintenance and repairs. In addition, since the meter is on the primary side of the transformer, the consumer pays for the transformer losses.

For this utility, on-peak hours last from 8 a.m. to 8 p.m., Monday through Friday. All other hours, including weekends, are off-peak.

Demand on this bill is the maximum, 15-min integrated kW demand created during on-peak hours. The bill doesn't say so, but the utility measures demand using a "sliding window" calculation; that is, the consumer can't synchronize with the utility's time interval to hide high-kW, short-time loads.

Extra kVA is determined by dividing the maximum demand by the average power factor for the month.

The **service charge** is authorized by the public utility commission (PUC), to cover some of the utility's fixed costs.

Energy adjustment charges were instituted in the 1970s, when fuel costs were unstable. An educated guess is made for the cost of fuel for a 12-month period and then reflected on the bill. If this guess is off by more than $20,000,000 during the 12-month period, the PUC makes a correction, which is later reflected in the bill. During stable times this usually results in a credit, but if the guess is more than $20,000,000 too *low*, it can add to the cost of power.

This utility also has a curtailable program. When requested, the customer agrees to shed at least 500 kW. The customer is then paid around $4.50 per kW of load shed. This works out well for organizations that can knock off a block of power for a while, or for plants that have on-site back-up power. Back-up power needs to be tested periodically anyway, so it makes sense to run it during peak periods and get reimbursed for it.

POWER FACTOR

Chapter 6

THE BEER ANALOGY

The power factor concept isn't as complicated as you may think. Let's draw an analogy to a mug of beer. Beer is the product that you want, and foam is a less-desirable by-product. Everyone understands that there will always be some foam with the beer; for the bar and restaurant industry, the foam is considered a cost of doing business, Figure 6-1.

Two types of power travel over power lines: **real power** (the beer) and **reactive power** (the foam). The combination of these is **apparent power** (the contents of the entire mug).

Real power is the easiest to explain. By definition, power is the ability to do work. It is expressed in watts (W).

An oversimplified explanation is that real power is produced when the voltage and current are traveling in the same direction.

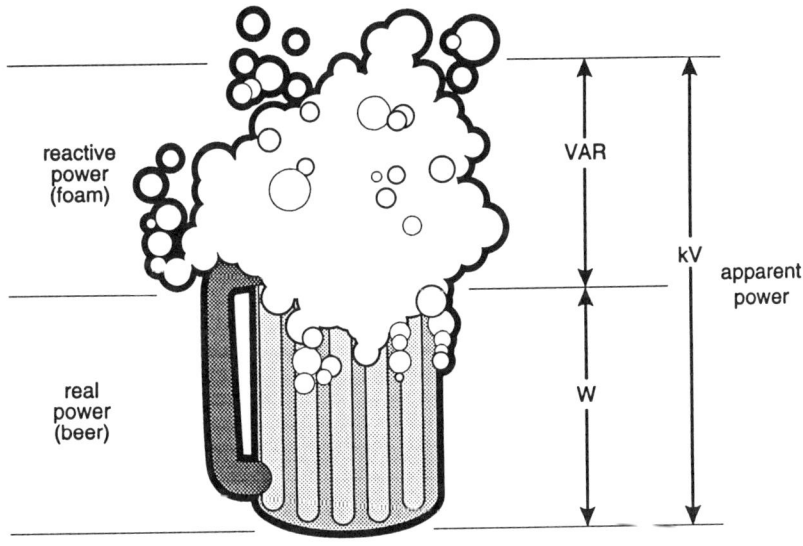

Figure 6-1. The power factor beer analogy

Volts times amps equals power. When a positive times a positive or a negative times a negative are multiplied together, the result is a positive answer.

Reactive power is not as straightforward. Expressed in volt-amps-reactive (VAR), it's also referred to as magnetizing, phantom, and/or negative power. To follow through with the oversimplified explanation, this power occurs when the voltage and current are traveling in different directions. Again, volts times amps equals power, but when a positive and a negative are multiplied together, the result is a negative.

Although the simple analogy of the beer mug shows that the beer and foam add directly on top of each other, the power types they represent actually go off in different directions. If you look ahead to Figure 6-5, you will see that real and reactive power go off at a 90-degree angle from each other. Apparent power, the combination of real and reactive power, is the third leg in this "power triangle." Apparent power determines the heating effect on ac equipment and systems. Expressed in kV, all elements of the power systems must be sized to accommodate the heating effect of the kV load.

(Keep in mind that reactive and apparent power occur in ac systems only, not in dc systems.)

Much the same way that too much foam takes up room in the beer mug, excess reactive power takes up room in the distribution system. By correcting the power factor, more real power capacity is available. This released capacity can save you from needing to run a new cable, or even from installing an additional transformer.

THE COMPLICATED THEORY: RESISTANCE, INDUCTANCE, AND CAPACITANCE

If you understand the beer metaphor, chalk this chapter up as a success. However, the ac theory behind it is a bit more complicated.

Dc theory is not as hard to understand as ac theory. It's simple and clear-cut, and all you have to deal with are volts, amps, and ohms. Unpleasant topics like power factor are not an issue with dc theory, because it's all beer and no foam! Ac theory, on the other hand, makes us talk about phenomenon like inductance and lagging currents, capacitance and leading currents, and phase angles — all of which tend to muddy the water.

All of these things happen because of the nature of alternating current. The higher the frequency, the more pronounced these occurrences become. Although 60 Hz is considered a low frequency, these occurrences still exert some influence and must be taken into account when figuring efficiencies and the cost of electricity consumption.

To get started, let's take a look at resistance, inductance, and capacitance.

Resistance is the simplest concept, because inductance and capacitance do not figure into the picture. So, to use another analogy, imagine how water loses pressure as it flows through a pipe; likewise, electric current drops voltage as it flows through a conductor. In both cases, energy is dissipated in the form of heat — the result of friction. The electrical symbol for resistance is —/\/\/\—.

Plug in a load that is almost purely resistive, such as an incandescent light bulb or a toaster. When voltage and current are graphed, as in Figure 6-2, we see that they

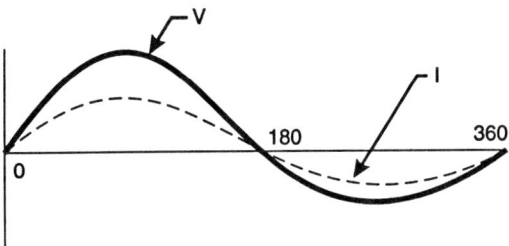

Figure 6-2. Voltage and current in phase

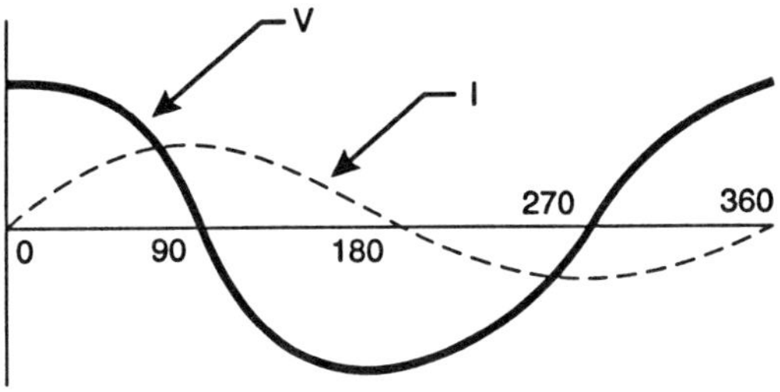

Figure 6-3. A purely inductive load

reach their minimums and maximums together and cross zero together. This is called being **in phase**.

Inductors restrict changes in current flow. In high school physics we're taught about inertia: Objects at rest tend to stay at rest; objects in motion tend to stay in motion. This theory is applied in the way inductors affect current flow. When power is first applied across an inductor, the current is zero. At that instant, the inductor acts like an open circuit. A magnetic field builds up and allows current to flow, but the current is always 90 degrees behind the voltage and it can't catch up. Not surprisingly, the electrical symbol for an inductor is a spring.

Now plug in a load that is almost purely inductive, like a length of wire wound around a soft iron core. Figure 6-3, another voltage and current graph, looks different; current is maximum when voltage is zero, and voltage is maximum when current is zero. Look at when the voltage and current pass through zero, and note that current lags voltage by 90 degrees. This is characteristic of a purely inductive load.

Capacitors resist changes in voltage. What inductors do for current, capacitors do to voltage. When power is first applied to a capacitor, the current is at its maximum value and voltage is zero. At the instant power is applied, the capacitor acts like a short circuit. A charge eventually builds up, allowing the voltage to reach its maximum level. But the current is always 90 degrees ahead of the voltage and never falls behind. The electrical symbol for a capacitor is ⊣⊢.

The voltage and current graph of a "perfect" capacitor is similar to the inductor's graph: When current is maximum, the voltage is zero and when voltage is maximum, the current is zero, Figure 6-4. The only difference is that, for the capacitor's graph, the current leads voltage by 90 degrees. This is characteristic of a purely capacitive load.

Now let's find out what all of this has to do with power factor.

POWER FACTOR DEFINED

Power factor is the measurement of the time phase difference between the voltage (E) and the current (I) in an ac circuit. It is represented by the cosine of this phase difference. (The cosine of the phase angle (ϕ) is multiplied by 100 and is expressed as a percentage.) To illustrate this, let's use a phase angle difference of 90 degrees. The cosine of a 90-degree angle is 0; therefore, the power factor is 0% (cosine = 0,

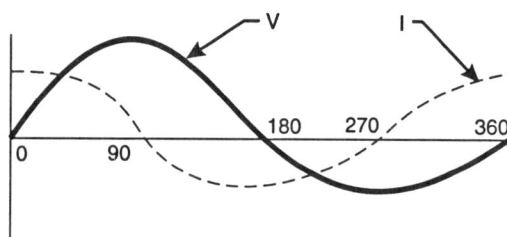

Figure 6-4. A purely capacitive load

0 x 100 = 0% power factor). If the phase angle were 60 degrees, the cosine of which is 0.500, the power factor would be 50% (cosine = 0.500, 0.500 x 100 = 50% power factor). This is true whether current lags or leads the voltage.

The formula used to determine power factor is as follows:

$$\text{Power factor} = \frac{\text{true power (P)}}{\text{apparent power (E} \times \text{I)}}$$

$$= \frac{P}{E \times I}$$

$$= \frac{kW}{kVA}$$

Where: P = power in watts (W)

I = the current in amperes (A)

E = the potential difference in volts (V)

Cos φ = the cosine of the phase angle

Notice that the power factor can never be greater than one. Engineers refer to this as unity. For example, a single-phase motor draws 4.5 A at 220 V. A wattmeter reads 702 W. What is the power factor?

$$\text{Power factor} = \cos \phi = \frac{720 \text{ W}}{220 \text{ V} \times 4.5 \text{ A}} = 0.728$$

What is the phase angle?

Power factor = cos φ = 0.728. Find the cosine column in a trig. table. The angle that corresponds to 0.728 is equal to 43.31 degrees. (The arc cosine function on a calculator yields the same result. Arc cosine means "the angle whose cosine is.") Therefore, the phase angle is 43.31 degrees.

Whenever there is talk about power factor, triangles show up, Figure 6-5. They graphically show that kVAR runs at a 90-degree angle to kW; that capacitive reactance cancels out inductive reactance; and that kVA is the result of kW and kVAR.

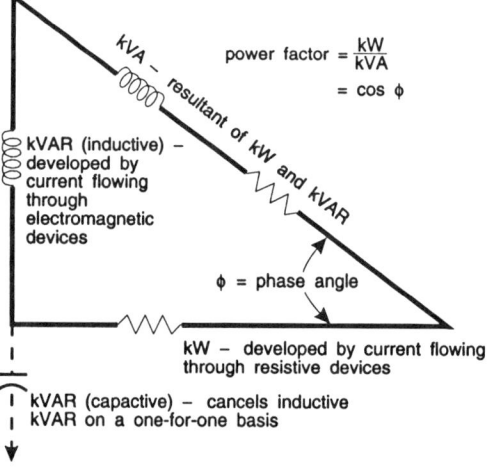

Figure 6-5. KVAR running at a 90-degree angle to kW

Power companies don't like excessive reactive power because it generates heat and uses up capacity in equipment and lines. Their physical plant, transformers, and distribution lines need to be sized for both the "beer" and the "foam." Public utility commissions let power companies charge for excessive power factor by either averaging it out and charging all customers, or by charging a penalty to customers with poor power factors. Some power companies charge by kVA instead of kW, because kVA includes both real and reactive power.

IMPROVING POWER FACTOR

Poor power factor also hurts building performance. For one thing, voltage regulation is decreased for a transformer supplying a load with a low power factor. Distribution lines also can accept additional capacity if their power factor is improved. By getting rid of the "foam," additional power can be delivered without adding another transformer or running a larger line.

The most common way to improve power factor is to add capacitors. The reactive power is then drawn from the capacitors instead of the power lines.

To reduce the power factor penalty from the electric bill, capacitors are installed at the service entrance. Figures 6-6a and 6-6b compare a partially loaded motor's power factor, both with and without a capacitor. To improve the plant's distribution system, capacitors are placed at the power-consuming equipment.

To understand power factor theory, we should become familiar with terms like "inphase resistive loads," "leading capacitive loads," and "lagging inductive loads."

To calculate power from the graphs in Figures 6-7, 6-8, and 6-9, we algebraically multiply voltage times current. (Algebraically means that you pay attention to the signs. A positive times a positive and a negative times a negative both yield positive numbers. However, a positive times a negative yields a negative number.) In this case, a positive number means power is flowing toward the load, and a negative number means power is flowing away from the load.

In the purely resistive load graph, Figure 6-7, the voltage and current always are in conjunction, or in phase. The multiplication of voltages and currents yields positive numbers. Power always flows toward the load.

In the pure inductive and capacitive load graphs, Figures 6-8 and 6-9, multiplication of the voltages and currents yield numbers in which half are positive and half are negative. When totalled, they cancel out to zero. *There is no real power in a purely inductive or capacitive load.*

As we are constantly reminded, nothing is perfect. All loads contain resistive, inductive, and capacitive components. In loads such as motors and lamp ballasts, the inductance cancels out what little capacitance exists. Figure 6-10 shows a real-world situation where the load is mostly resistive, but current lags behind the voltage, producing some reactive power.

Notice the symbol ϕ in the Figure 6-10: ϕ represents the phase angle, or the difference between where the voltage and current curves cross zero.

How can we improve power factor? As stated earlier, you add capacitance to cancel out some of the inductance, thus improving power factor. First, an engineering evaluation must be performed to determine where and how many capacitors are to be added, and what type of over-current protection needs to be installed.

Power Technology

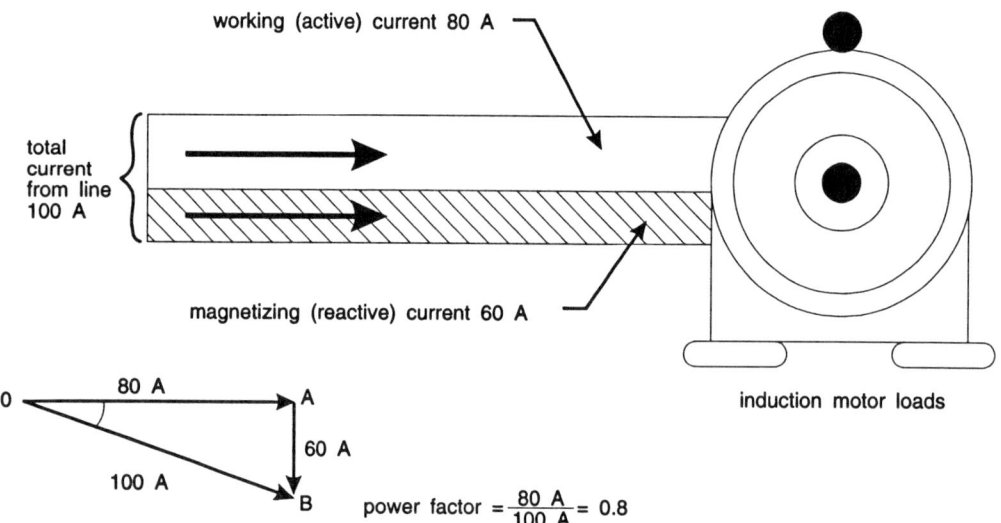

Figure 6-6a. Partially loaded motor without capacitor

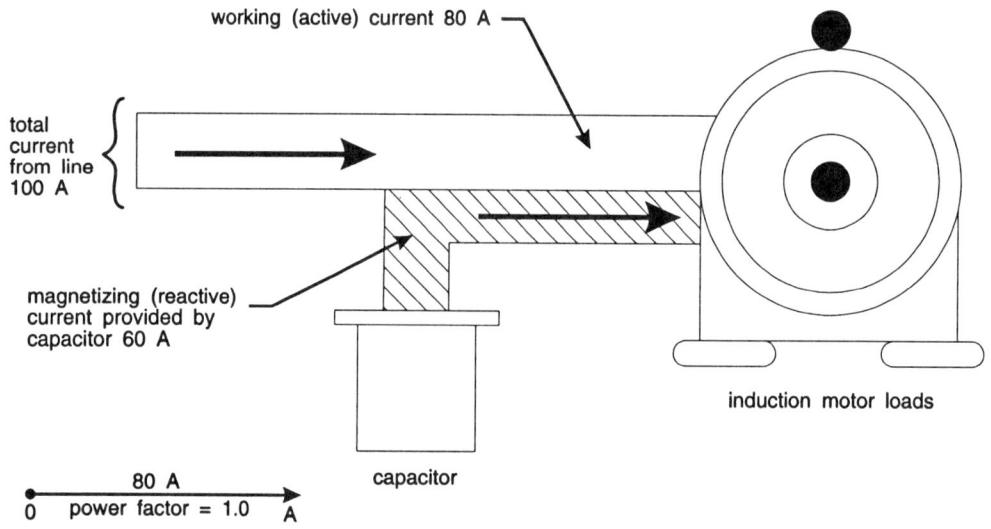

Figure 6-6b. Partially loaded motor with capacitor

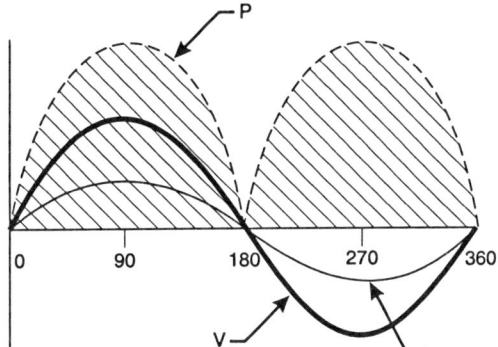

Figure 6-7. Purely resistive load

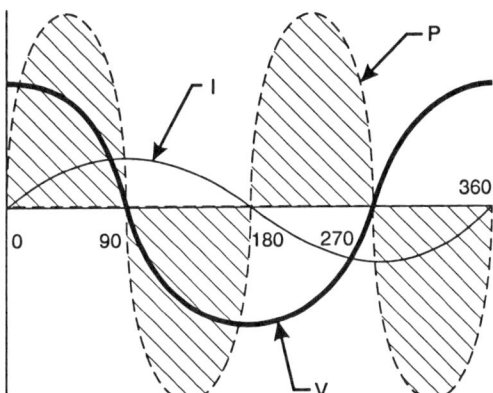

Figure 6-8. Purely inductive load

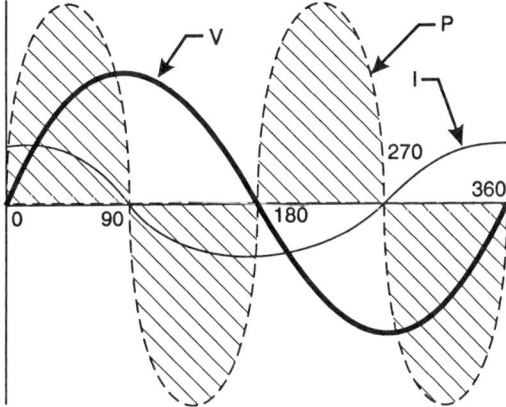

Figure 6-9. Purely capacitive load

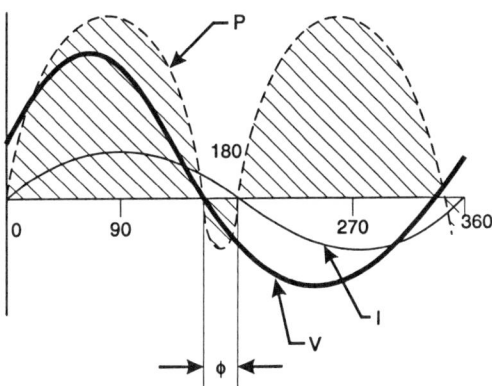

Figure 6-10. Real-world situation

Note: *Too much of a good thing can be bad. If you get a plant's power factor over 95%, motors can become overexcited (electrically speaking). Over excitation results in high transient voltages, currents, and torques, thus increasing safety hazards to personnel and causing possible damage the motor or driven equipment.*

For motors, the best place to install capacitors is at position No. 2, which is shown in Figure 6-11. The current passing through the motor overload is not changed. Position No. 3 is good too, but because the current going through the overloads is decreased, the overloads need to be changed.

Under some circumstances, capacitors cause system problems. Don't connect power factor correction capacitors at the motor terminals of elevator motors; multi-speed motors; plugging or jogging applications; open transition, wye-delta, autotransformer starting; and some part-winding motors. If power correction is desired, connect the capacitors upstream of the load, such as position No. 1.

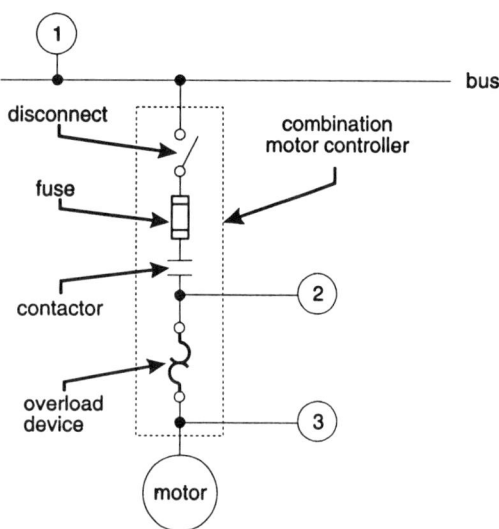

Figure 6-11. Three possible positions at which to install motor capacitors

chronous motor. Its advantages include no voltage spikes when changing settings and absorbing low-level harmonics.

Another concern is, what happens when capacitors are switched on and off from the system. Power companies have capacitors installed in the field to help with their distribution. Switching them off-line can cause voltage spikes. Spikes went unnoticed for years, but with computer equipment becoming more common, these spikes can cause problems. Also, adding capacitors deserves additional consideration if excessive harmonics are present (more about harmonics in a later chapter). In short, engineering assistance usually is necessary when adding capacitors.

Table 6-1 shows a method for determining the size of capacitors required.

Another option for correcting poor power factor in large installations is synchronous motors. These motors can be excited so that they have a leading power factor, thereby canceling-out some of the system's inductance. Yet another variation of this is the synchronous condenser, a shaftless syn-

Power Factor

Original power factor (%)	Desired power factor (%)				
	100	95	90	85	80
60	1.333	1.004	0.849	0.713	0.583
62	1.266	0.937	0.782	0.646	0.516
64	1.201	0.872	0.717	0.581	0.451
66	1.138	0.809	0.654	0.518	0.388
68	1.078	0.749	0.594	0.458	0.328
70	1.02	0.691	0.536	0.4	0.27
72	0.9654	0.635	0.48	0.344	0.214
74	0.909	0.58	0.425	0.289	0.159
76	0.855	0.526	0.371	0.235	0.105
77	0.829	0.5	0.345	0.209	0.079
78	0.802	0.473	0.318	0.182	0.052
79	0.776	0.447	0.292	0.156	0.026
80	0.75	0.421	0.266	0.13	
81	0.724	0.395	0.24	0.104	
82	0.698	0.369	0.214	0.078	
83	0.672	0.343	0.188	0.052	
84	0.646	0.317	0.162	0.206	
85	0.62	0.291	0.136		
86	0.593	0.264	0.109		
87	0.567	0.238	0.083		
88	0.54	0.211	0.056		
89	0.512	0.183	0.028		
90	0.484	0.155			
91	0.456	0.127			
92	0.426	0.097			
93	0.395	0.066			
94	0.363	0.034			
95	0.329				
96	0.292				
97	0.251				
99	0.143				

Assume total plant load is 100 kW at 60% power factor. Capacitor kVAR rating necessary to improve power factor to 80% is found by multiplying kW (100) by multiplier in table (0.583), which gives kVAR (58.3). Nearest standard rating (60 kVAR) should be recommended.

Table 6-1. Method for determining the size of capacitors

HARMONICS

LINEAR AND NON-LINEAR LOADS

Harmonics are currents or voltages that are multiples of the fundamental frequency. If, for example, the fundamental frequency is 60 Hz, the second harmonic would be 120 Hz, and the third would be 180 Hz.

Linear loads do not cause harmonic problems. A linear load draws current in direct proportion, and in a smooth, sinusoidal fashion, to the applied voltage. Sinusoidal means in the shape of a sine wave. Heating elements and motors create linear loads. Before the proliferation of electronics, linear loads made up the majority of building loads.

Non-linear loads, as graphed in Figure 7-1, draw current in abrupt pulses. These pulses cause harmonic currents which, in turn, result in voltage distortion and ultimately cause more current harmonics in other parts of the power system.

Switching-mode-type power supplies are efficient, small, and do a great job of satisfying the requirements of modern electronics. The problem is, switching-mode power supplies use non-linear loads and generate harmonics in the facility's electrical distribution system.

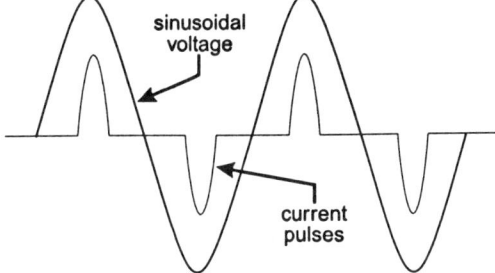

Figure 7-1. Non-linear load

Some equipment loads that switch current abruptly and cause harmonics include power supplies for computers, fax machines, copiers, variable-frequency motor drives, UPS machines, solid-state heating controls, and electronic fluorescent lighting ballasts.

The current pulses caused by switching-mode-type power supplies can be compared to the output of a reciprocating compressor. During most of the piston's travel, no air is pushed out of the cylinder because it isn't compressed enough to overcome the receiver pressure. Only when it's nearing the end of its stroke is enough pressure developed so that air is discharged. The discharge airflow occurs in relatively short pulses, very similar to the current flow in switched power supplies.

Figure 7-2 shows how these current pulses, or blips, distort the voltage as well as the current wave form, and how they can make sensitive data-handling equipment miss a beat now and then. Therefore, it is critical to connect sensitive loads to **clean lines**. (More on these later in the chapter.)

Let's take a minute to put this information into perspective. Just because a new fax machine is installed next to a personal computer doesn't mean the facility's power distribution system is going to crash. It *does* mean that the cumulative effect of running all this electronic equipment will have an effect on power quality.

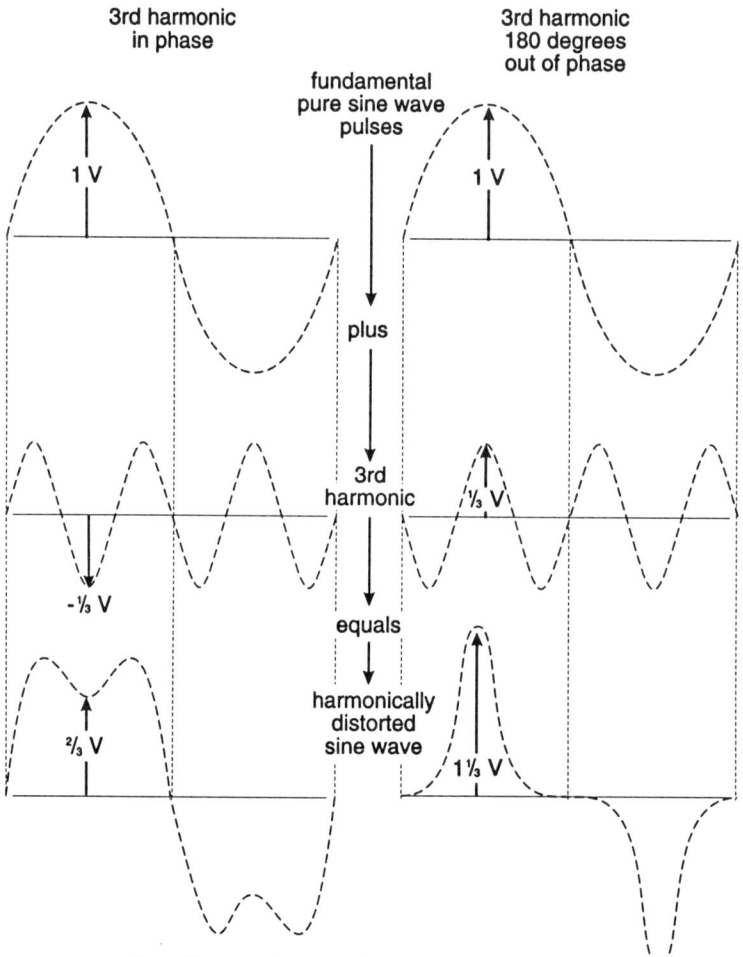

Figure 7-2. How current pulses distort voltage and current wave form

HARMONIC SYMPTOMS, MEASUREMENTS, AND CURES

SYMPTOMS

Harmonic troubles usually show up first in the form of overheated neutrals and transformers. Balanced three-phase linear loads are depicted in Figure 7-3. The loads on hot legs A, B, and C are all connected to the common neutral. The three illustrations are all of the common neutral, but show the contributions from an individual phase. When all the unshaded wave forms are added up, they cancel each other out. Only the harmonic **triplens** combine to produce a current. For simplification, only the fundamental and third harmonic are shown. However, the 9th, 15th, 21st, etc., harmonics add up the same way. All the other harmonics cancel themselves out.

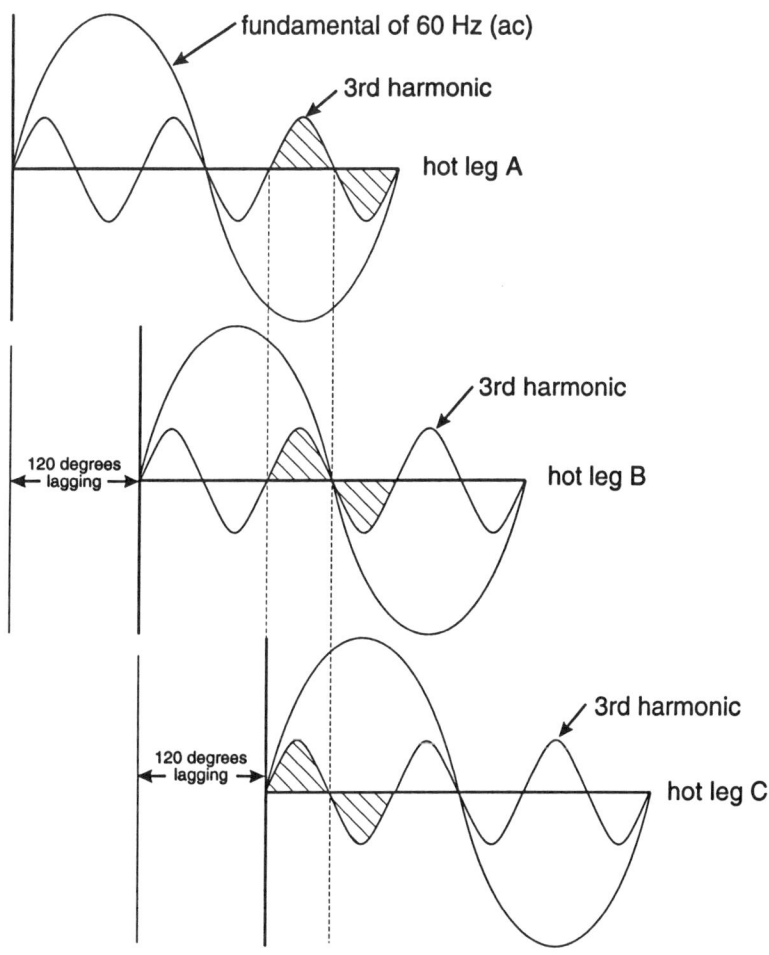

Figure 7-3. Three-phase load with harmonic activity

Another symptom of non-linear harmonic problems is circuit breakers tripping or fuses blowing far below their rated capacity. The distorted wave forms generated by the odd harmonics are fooling these devices into tripping early.

MEASUREMENTS

Unless you have the proper diagnostic instruments, the problem of harmonics can drive you crazy. Standard meters actually sense the average value of a sine wave. The meter scale or digital readout is calibrated to read RMS (root mean square) values. Unless otherwise stated, all values of ac voltage and current are RMS values.

RMS values are used to directly compare the heating value of ac current to dc current. One ampere of RMS ac flowing through a given resistance produces heat at the same rate as a dc ampere. Figure 7-4a shows the relationship between peak, RMS, and average values of a sine wave.

Figure 7-4b shows how harmonics distort a sine wave. Because of this distortion, the values in Figure 7-4a, and the meter calibrations of averaging instruments, are no longer valid. A true RMS meter measures RMS values directly; its accuracy is not affected by distorted wave forms. True RMS meters are more expensive than averaging meters.

For example, if you used a standard averaging ammeter to measure the input current of a typical personal computer, you would get a reading of 1.35 A. A true RMS meter would give a reading of 2.15 A. The greater the distortion, the greater the difference between readings from the two types of meters. (An averaging meter typically reads 20% to 50% lower than a true RMS meter.)

The current-carrying capability of conductors is based on the heating effect of a current's RMS value. Averaging meters on distorted waves can give a false lower value, and conductors could overheat due to inaccurate information.

Harmonic analyzers, although they are even more costly than RMS instruments, are capable of sorting out and measuring the am-

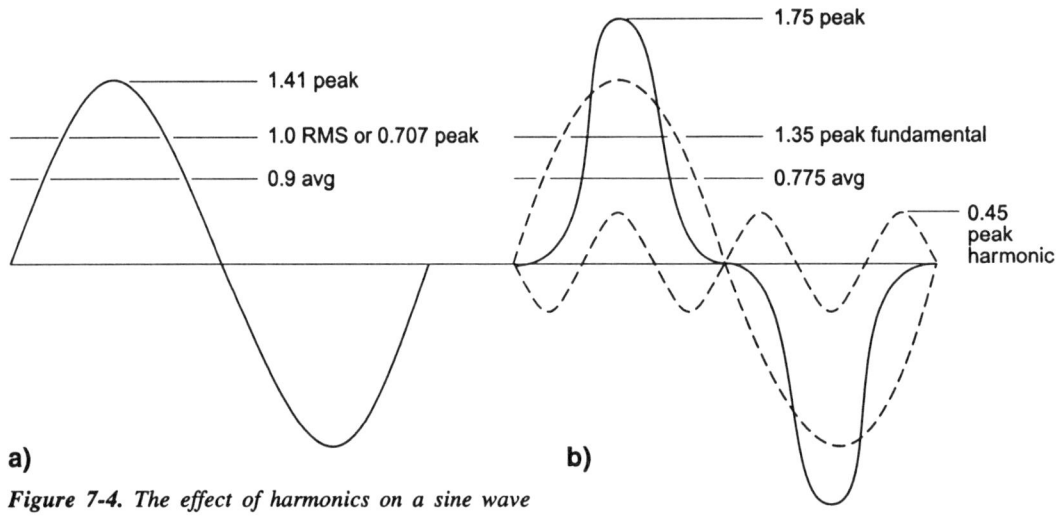

Figure 7-4. The effect of harmonics on a sine wave

plitude, or strength, of all the frequencies on the power line. They usually measure in percentage form compared to fundamental frequency. (Measuring instruments are discussed further in Chapter 12.)

CURES

Case #1: The efficiency of equipment designed for 60-Hz operation is hurt when higher-frequency harmonic currents are applied. Transformers are especially sensitive to harmonics. The inefficiency shows up as additional heat.

There are a couple of ways to reduce transformer problems. The quick-and-dirty way is to derate the transformer. Derate means to operate a piece of equipment below its normal rating. For example, a 50-kVA transformer would only be loaded to 25 or 35 kVA. If that doesn't solve the problem, then look around for a better transformer. One way transformers are rated is by their temperature rise. Standard transformers are rated at 150°C rise. The premium models have 115°C and 80°C rises. These are built better to run cooler, and often this is all it takes to handle the harmonics.

If the problem is severe, then go with the super-premium "K"-rated transformers. This "K" factor is a calculated numerical rating that indicates the severity of the harmonic problem. Data for the K calculation is obtained from harmonic analyzers.

In conjunction with installing better transformers, don't forget to increase the size of the neutral to handle the increased heat created by the harmonics.

Case #2: Between the transformer and equipment is an electrical panel. With standard panels, the neutral bus bar and its connections are likely to develop hot spots due to excessive neutral current.

They also might mechanically resonate — hum — at the higher harmonic frequencies, for which they were not designed. But don't despair; panel boards now are available with oversized neutral buses, which reduce heating and provide extra stiffening to lower the noise.

Case #3: Voltage harmonics cause problems in induction motors as well. Like transformers, the eddy current losses are greater at higher frequencies. In addition, some harmonics, notably the 5th, have a "negative sequence"; honestly, they seem to be trying to run the motor backwards. Total harmonic distortion for motors should be less than 5%.

CLEAN LINES, ETC.

Remember, back near the beginning of the chapter, when I said we'd talk "later" about how to get clean lines? Well, "later" is now. I'll also discuss how to eliminate overheating.

A lot of design options can improve a harmonic situation (hence cleaning up the power transmitted in the lines), without the expense of purchasing new K-rated equipment. For starters, you could run a dedicated line to each piece of equipment.

Some delicate equipment, like electronic data apparatus (computers, phone systems, etc.), are very sensitive to power line noise from other loads. One way to minimize undesirable power line noise is to put that piece of equipment on its own power line.

When an electrician or engineer states that a piece of equipment is on a dedicated line, they mean that piece of equipment has exclusive use of the particular circuit. There are no other loads on that line to induce noise or cause trouble.

It might not seem right to have a dedicated 20-A line serving one 4-A load, but this method goes a long way toward achieving clean power. It may use up a lot of panel space, but with high neutral currents the panel would have to be derated anyway. With this dedicated arrangement, your old standard panels will do just fine.

Another strategy you can use is to run each feed separately and not in wire troughs. Wire troughs may bring some order to the mess under raised floors, but they can be a big a source of trouble. Just think: all those wires with their dirty power running parallel in close proximity; harmonics could spread from one circuit to another like the plague.

Another problem with troughs is that they can become overfilled, possibly causing problems and expense for the building owner. There is enough space for 615 #12 THHN conductors in a 6- by 6-in. wire trough; when a trough holds more than 30 conductors, or is 20% full, the Code requires derating to prevent heating problems.

Remember, the neutral is a current-carrying conductor. By running separate cables, these problems are avoided. Also, try to keep power cable runs as short as possible, by placing the distribution panels close to the load.

Filters are another option to use under severe conditions. Like oil or air filters, which are designed to block certain-sized particles while allowing others to pass, electronic filters can be designed to block higher harmonic frequencies, while allowing 60-Hz power to pass through. Application of filters must be done by someone with the knowledge and experience for this design situation.

Remember, the more severe the problem, the more engineering assistance is required.

Finally, don't overlook the obvious, like making sure all connections are tight, and redistributing loads to balance load currents.

POWER LINE DISTURBANCES

POWER PROBLEMS

A total power failure in the computer center would be disastrous; however, such extreme cases make up a very small percentage of power problems. Voltage dips, power surges, and brief power failures also disrupt sensitive electronic equipment operation, such as computers. The old Bell System studied power line disturbances in the 1970s and came up with these findings.

Out of all power line disturbances studied, voltage dips accounted for the highest percentage. The distribution of disturbances was as follows:

Voltage dips — 87%

Impulses — 7.5%

Power failures — 4.7%

Power surges — 0.8%

Critical sensitive loads, mainly computers, require clean, constant power for proper operation. The problem is, disturbances aren't always apparent — except that they cause data loss and erratic operation.

This chapter isn't intended to make you an expert on computers and their power supply, but to give you an additional troubleshooting tool: If you suspect power line disturbances are disrupting operation of a building's comfort control system, you can check for symptoms cited here.

Throughout the chapter, keep in mind that a lot of power problems originate in the building, not in the outdoor power lines or at the power plant. Most utilities will help you determine if the problem is on their side of the meter or yours.

The severity of the problem determines what corrective action to take. To determine just how severe the problem is, use one of many instruments on the market that can measure and analyze the extent of power line disturbances. Prices may vary, but the good ones are rarely inexpensive.

Note: *If you or your staff are not familiar with this apparatus, you would be better off to have a consultant provide the expertise and equipment.*

Line interference, dedicated circuits, isolated grounds, and transformers were discussed in a previous chapter. Let's review them briefly.

MUTUAL LINE INTERFERENCE

Often power problems can be solved by proper wiring methods. Electrical wiring is frequently run together in troughs under the raised computer floor. Computers generate harmonics, and the resulting line disturbances easily pass from one circuit to another. A simple solution is to run each circuit separately from its distribution panel. This might increase the cable disorder under the floor, but it will clean up the power.

DEDICATED CIRCUITS

Another temptation is to put more than one piece of equipment on the same circuit, especially equipment that draws very little power. A piece of equipment that generates a large amount of harmonics can overwhelm a smaller piece of equipment on the same circuit.

Putting each piece of equipment on its own dedicated circuit helps put all the equipment on an even footing. This does eat up a lot of panel space, but it guarantees better equipment operation.

ISOLATED GROUNDS

The **typical building ground** takes care of items such as outlet junction boxes and other noncurrent-carrying conductive surfaces, for which low-level circulating currents are not a problem. The **isolated ground** takes care of sensitive electronic equipment.

A lot of manufacturers specify an isolated ground for their equipment, to prevent ground loops. All grounding conductors *do not* carry the same voltage potential; there are, in fact, small voltage differences throughout the system.

Whenever there is a voltage difference, there is going to be current flow. These stray currents, or ground loops, have no impact whatsoever on most noncurrent-carrying equipment; however, ground loops can cause some sensitive computer equipment to act a little nuts.

An isolated grounding system eliminates these loops, bringing all of the equipment grounds back to a single grounding point using insulated wire. (To do this, and also to comply with the Code, a special outlet and two separate grounds must be used, as shown in Figure 8-1). They are allowed to pass through as many power panels as required to reach the single-point ground. The regular building ground, however, must be terminated in the first power panel it reaches. Isolated ground receptacles are usually orange in color.

On regular outlets, the ground sleeve is attached to mounting tabs that automatically bond together the junction box, enclosure, and equipment. On an isolated ground outlet, the ground sleeve is separate from the mounting tabs. So, when equipment is plugged into an isolated ground outlet, the equipment is separate from the common ground.

To comply with the National Electrical Code, equipment enclosures must be grounded. This ground can be made the conventional way, without elaborate precautions, since ground loops are not a problem with enclosures.

TRANSFORMERS AND DIRTY POWER

Sometimes the power supplying the computer room is considered "dirty," thus potentially dangerous to sensitive equipment. Transformers can help clean up incoming power.

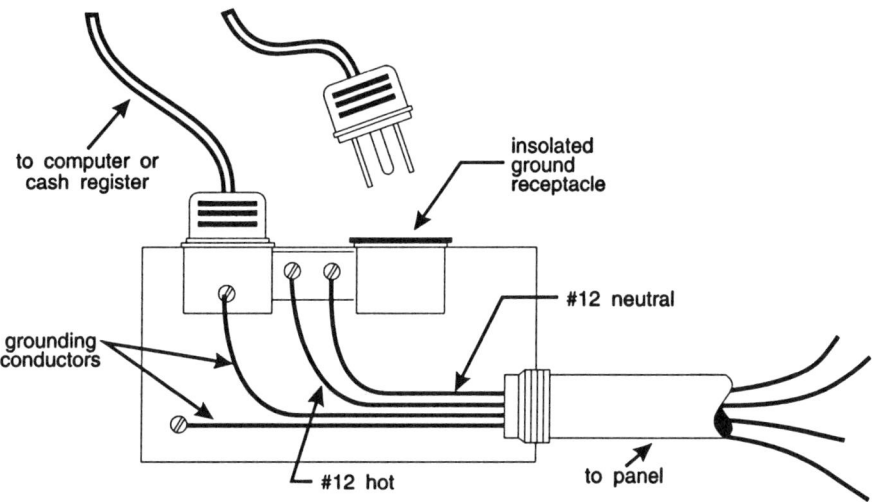

Figure 8-1. Isolated Ground

DIRT

Common-mode noise is the technical term for line-to-ground transients. They are troublesome because they bypass equipment power supply filters and easily penetrate sensitive electronic grounding systems. They often directly affect the critical low-level digital signals that are essential for proper computer operation.

Common-mode noise travels from the primary to the secondary of a transformer by means of electrostatic coupling. This coupling can be broken by placing a thin sheet of foil between the primary and secondary, effectively shorting out the spikes and noise.

Transverse-mode is the technical term for across-the-line disturbances or line noise. It actually contributes relatively little to equipment malfunction or failure.

POWER DISTRIBUTION UNITS

Many installations use power distribution units (PDUs) to improve power quality. These units consists of a high-quality isolation transformer, distribution panels, and instrumentation, that read incoming and outgoing volts, amps, and watts.

UNINTERRUPTIBLE POWER SUPPLIES

Even though utility power is good — that is, relatively clean — good isn't good enough when computers are concerned. Even utilities use uninterruptible power supply (UPS) systems to protect their computers.

Please note that a UPS system is definitely interruptible. When normal power fails, the battery energy in most systems is good for about 15 min. During that time it is expected that the critical load will be shut down in an orderly fashion, or a generator will be put on line.

There is a price to pay for this protection. In addition to the initial cost, periodic maintenance must be performed both on the UPS machine and the batteries. The UPS consumes power on its own, so the electric bill is higher; and, because this work is specialized, it often involves separate contracts from various vendors.

But when all is said and done, the penalty of lost computing time and data far outweighs the cost of UPS systems.

Large systems require dedicated space for the equipment, and special steps must be taken so battery acid doesn't harm the environment.

The two main types of UPS systems are **static** (the output is the result of solid-state devices with no moving parts) and **rotary** (the output is the result of a rotating motor-generator set). These are further subdivided into **on-line** and **off-line** designs.

STATIC UPS SYSTEMS

All static UPS systems have a rectifier (converts ac to dc), a set of batteries, an inverter (converts dc to ac), and an automatic transfer or manual bypass switch.

The **rectifier's** job is to keep the batteries charged. In the case of on-line machines, it also powers the inverter under normal conditions.

The **batteries** provide power during power failures and other aberrations.

The **inverter** converts the dc from either the rectifier or the batteries, and turns it into ac for the critical load.

The **transfer or bypass switch** connects or disconnects the UPS from the critical load.

Figure 8-2 shows a typical static UPS system. During normal operation the UPS supplies power to the load. The automatic transfer switch transfers the load to the normal power line in the event of a UPS malfunction, or if the UPS has to be taken off-line for maintenance. If the transfer switch is based on solid-state components, it is referred to as a **static transfer switch**.

Figure 8-3 shows the normal power line replaced by an emergency generator; and *this* is what makes an uninterruptible system truly uninterruptible. The generator is automatically started after a number of minutes into a power failure. When the system determines that the generator's output voltage is correct and in phase with inverter output, it assumes the load. With some systems, the generator's output is fed into the rectifier.

ROTARY UPS SYSTEMS

The heart of a rotary UPS system is a motor-generator set (often referred to as MG sets). Under normal conditions, the motor gets its power from the local utility. The critical load never sees dips, impulses, or surges, because it gets its power from the generator. Notice how a motor and its driven load takes a while to stop rotating after the power is turned off. This **flywheel effect**, caused by the weight of the MG set's rotating mass, is used to ride through short power dips. If the utility's power disturbance is under 50 millisec, the critical load won't sense the difference.

If the problem lasts longer, a battery-supplied inverter supplies the motor with power. The inverter can be less sophisticated; it only has to supply a less-demanding motor load, not a discriminating computer.

Some people think rotary systems are more reliable, because they don't have as many components as static systems. However, the generator delivers a pure sine wave to the

Power Line Disturbances

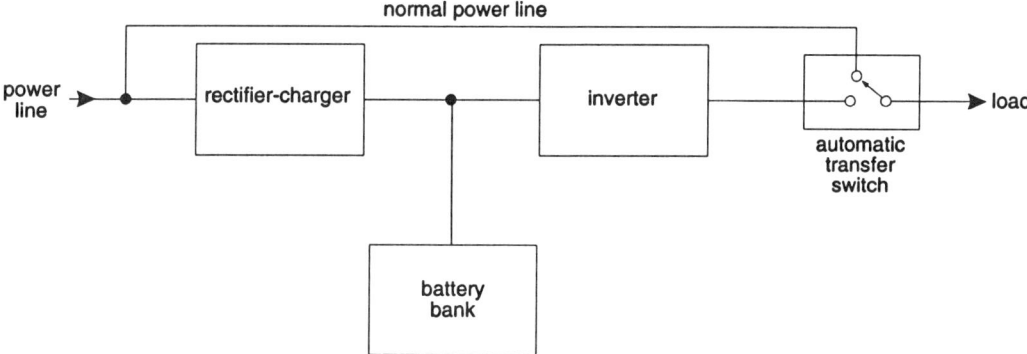

Figure 8-2. Typical static UPS system

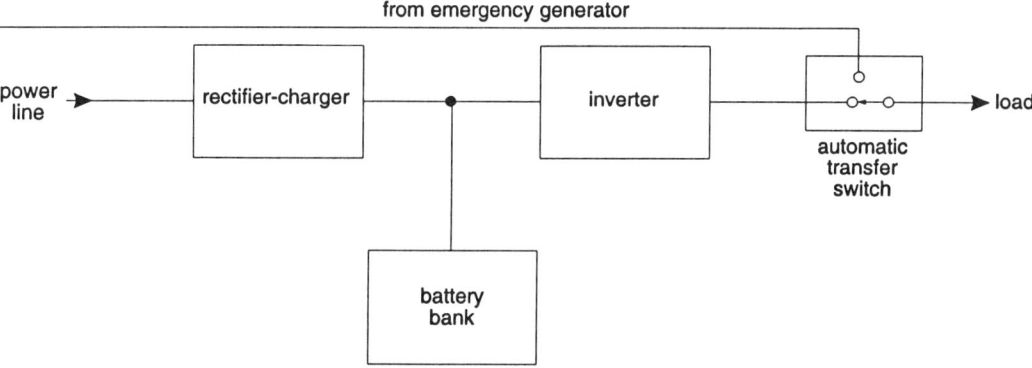

Figure 8-3. Normal power line replaced by an emergency generator

load. Bearings become the main maintenance item. Two disadvantages of rotary systems are:

1. They are more expensive.
2. They use more power than static systems.

One early-design rotary UPS system consisted of a generator, a flywheel, a motor, an electric clutch, and a diesel engine. Under normal conditions the utility supplied power to the motor, which turned the generator and flywheel. During a power failure the clutch engaged the diesel engine. The flywheel had enough stored energy to start the engine, which maintained power to the load.

ON-LINE AND OFF-LINE

The many types of UPS systems on the market can be divided into two general classes: off-line and on-line.

With an off-line UPS, the computer is fed directly from the utility's power line. The system quickly turns itself on when there is a power failure — within 50 millisec — in order not to lose any data. It doesn't run all the time, so it's less expensive to operate than the on-line type. An off-line UPS doesn't protect against spikes and dips unless power-conditioning equipment is added.

With a static on-line UPS, the inverter supplies power to the critical load; power aberrations never reach the load, because the UPS is between the incoming power line and the computer.

Chapter 9
LIGHTING TERMINOLOGY, SOURCES, AND BALLASTS

Light can account for up to 35% of your electricity bill. About 86¢ of each lighting dollar goes to energy. Every lighting dollar, in turn, saves up to 40¢ in cooling costs. Therefore, it's worthwhile to learn something about lighting terminology, sources, and basic technology.

TERMINOLOGY

Here are some terms that crop up frequently when discussing lighting.

Lumen output (lm) is how much total light is generated from a light source.

Watts (W) is a measure of electrical power commonly applied to lighting (i.e., 120-W lightbulb).

Efficacy is the lighting industry's terminology for **lumens per watt (lm/W)**. It describes how much visible light is emitted from a source per watt of power input.

Fixtures vs luminaires: For most people, a fixture is what a bulb is screwed into. For lighting professionals, a fixture is a white porcelain object found in bathrooms. What *we* call fixtures, *they* call luminaires.

Lamp mortality, or rated average life, is the measure of how many hours half of a group of lamps will remain lit. Every time a lamp is turned on its life is shortened; therefore, mortality tables are based on hours between starts.

Footcandle (fc) is the amount of light that makes it to an object's surface. (Technically, it's defined as the illuminance on a surface 1 sq ft in area on which there is a uniformly distributed flux of one lumen.) Many instruments on the market measure footcandles.

How much light actually makes it to the work surface depends on several factors. To deliver a satisfactory final product, lighting designers must take into account these variables:

- Luminaire efficiency — How much of the lamp's output makes it out of the fixture.

- Luminaire dirt depreciation — Clean luminaires deliver more light than dirty luminaires.

- Lamp lumen depreciation or lumen maintenance — As lamps age, their output decreases. Manufacturers supply tables and graphs for lumen depreciation.

- Ambient temperature — The effect of surrounding air temperature is considerable on the output of some lamps.

- Ballast factor — Different ballasts cause lamps to emit different levels of light. The proper ballast choice must be made if high light output and economy of operation are desired.

- Color-rendering index — This is a comparison, from 0 to 100, of the spectrum of light emitted by a lamp as compared with the sun. The higher the index, the better the appearance of objects illuminated by the lamp.

- Color temperature — When you pass electricity through a piece of metal, the electricity causes the metal to heat up until it begins to glow. At first it is red-orange. As the temperature increases, it becomes orange, then yellow, and eventually blue or blue-white. A fluorescent lamp with a color temperature of 4,100 Kelvin (K) is similar in color to a piece of metal heated to 4,100 K.

LIGHT SOURCES

With these terms under our belt, let's quickly look at a few types of light sources. (Figure 9-1 shows these sources rated according to their efficacy.)

Incandescent light, including quartz halogen, is at the bottom of the efficacy ladder. With very few exceptions, there is really no good reason to use incandescent lighting in commercial-industrial facilities. Any energy efficiency program should aggressively try to avoid these dinosaurs.

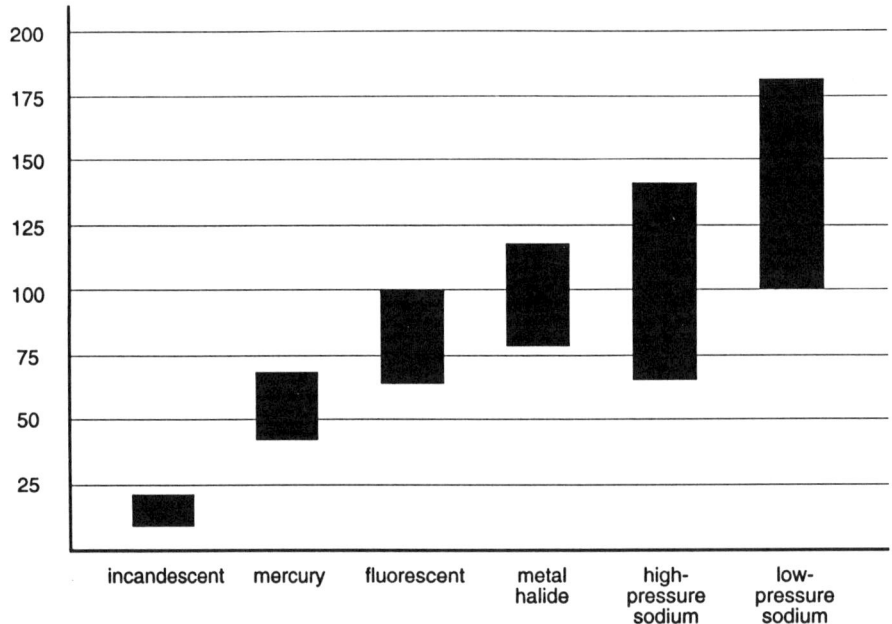

Figure 9-1. Approximate initial efficacies for a range of commonly used lamps (lm/W)

Edison started the electric light industry with the incandescent bulb. It produces light by heating a filament until it glows white hot — and therein lies its relative inefficiency.

Because it needs to get very hot before it starts to emit any significant light, 90% to 95% of its input energy is wasted as heat. Also, because the tungsten filament molecules are burned off, lamp life is short (typically 2,000 hours). Finally, these burned-off filament molecules blacken the bulb's surface, so light output decreases. The efficacy of incandescent bulbs is poor, approximately 10 to 20 lm/W. The color rendering index, however, is excellent, usually well over 90%. This makes incandescent light popular in retail applications, where it is important to show off a product in its best light.

Halogen gas can be added to keep an incandescent bulb from darkening, by recycling the tungsten deposited on the bulb wall back to the filament. This also yields a moderate lamp life increase over regular incandescents. Quartz glass, and sometimes special fixtures, need to be used because the bulb surface reaches such high temperatures. (Because quartz glass is used, this lighting source is sometimes referred to as quartz lamps.) There is an ultraviolet hazard associated with these halogen-incandescent bulbs. Efficacy and color rendering index are the same as regular incandescents.

Note: *Use gloves or a rag when handling high-performance incandescent halogen or projector bulbs. Dirt and oil from your hand causes localized thermal stress on the glass, possibly leading to premature failure.*

High-Intensity Discharge (HID) lamps are used outdoors (street and parking lot lighting) and indoors (warehouse and industrial) because of their high light output and efficiency. Mercury vapor, metal halide, low-pressure sodium, and high-pressure sodium are examples of HID lighting.

Mercury vapor was the first efficacy improvement over incandescent lighting. Although a big advancement for its day, this technology has seen its day.

Mercury vapor was the first popular high-intensity lamp to make it big replacing incandescent street lamps. Since their introduction in 1934, they have been improved to the point where efficacies are now up to 65 lm/W, and their life ratings exceed 24,000 hours. At first their color rendering was poor, but improvements in phosphor technology have made them acceptable for all but the most critical applications.

Metal halide lamps are, in effect, souped-up mercury vapor lamps. Their efficacies go up to 125 lm/W. Because they have a good color rendering index (50 to 60), they are suitable for applications where high light output and good color are required. A disadvantage of metal halide lamps is that they have a relatively long restart or restrike time (8 to 15 min) when the lamp is turned off momentarily.

High-pressure sodium lamps are popular because of their high efficacies (140 lm/W) and color rendition with newer designs. They have a short restart time (2 to 3 min) compared to mercury and metal halide lamps.

Low-pressure sodium lamps: The good news here is that efficacies range up to 183 lm/W. The bad news is that, since it is a monochromatic source, a low-pressure sodium lamp provides light of only one color (yellow), and is the absolute worst when it comes to color rendition. You can't recognize your own red, green, or blue car parked under this light source; it all looks gray.

However, they do restrike instantly when hot.

Fluorescent and low-pressure sodium are low-intensity light sources packaged in long tubes. They do not lend themselves to applications where a high degree of optical control is necessary. Mercury, metal-halide, and high-pressure sodium are point sources: They can be mounted in fixtures and aimed.

Fluorescent lighting was introduced commercially in the 1930s. It is the most common office lighting system. For years it was stuck on an energy plateau, but newer technology, such as electronic ballasts and improved lamps, are making this system more efficient.

Fluorescent lighting has been the mainstay of commercial lighting for years. In fact, the cool white and warm white T-12 lamps have been around since the 40s. (The 12 in T-12 is the diameter of the fluorescent in $1/8$-in. increments. A T-12, thus, is $1\frac{1}{2}$ in. dia. The newer T-8 lamps are 1 in. dia.)

Unlike incandescent bulbs, which convert energy to heat in order to produce light, fluorescent lamps convert energy to light by using an electric charge to "excite" a small amount of mercury vapor atoms, producing ultraviolet energy in the fluorescent tube. Although the human eye can't see ultraviolet light, this gaseous discharge causes the phosphor coating on the inside of the tube to fluoresce, thus emitting visible light.

The newer phosphor blends that are replacing cool white and warm white have increased the color rendering index from around 60 to over 90. Because little heat is generated, fluorescent lighting is more energy efficient than incandescent; fluorescent efficacies are greater than 80 lm/W.

Table 9-1 shows a mortality chart for fluorescent lamps. An argument against turning off lights has been that, every time a lamp is restarted — whether incandescent or fluorescent — its life is shortened. True, but remember that energy takes about 86% of total system lighting costs. (Labor is roughly 11%, and the lamp is 3%.) As a rule, if a space will remain unoccupied for more than 15 min, fluorescent lights should be turned off.

	Hours per Start					
Lamp type	3	6	10	12	18	Continuous
40 W preheat	15,000	17,500	21,250	22,500	25,000	28,125
40 W rapid start	20,000+	24,420	27,750	28,860	31,600	37,700
High output (HO)	12,000	14,000	17,000	18,000	20,000	22,500
Very high output (VHO)	10,000	12,500	14,990	15,980	17,980	24,980
Slimline (96T12)	12,000	14,000	17,000	18,000	20,000	22,500

Table 9-1. Mortality chart for fluorescent lamps

BALLASTS: MANUAL-, INSTANT-, AND RAPID-START SYSTEMS

The rest of the light sources are "discharge" devices that use ionized gas instead of a glowing filament to produce light. All discharge lighting requires a device called a ballast in order to operate. Ballasts are connected between the incoming power and the lamp. Ballasts:

- Provide current to cathode heaters;
- Provide the energy to ionize the gas;
- Limit current traveling through the tube to a safe level after the gas is ionized.

Notice that there are no voltage ratings for fluorescent tubes or other discharge-type lamps, as there are for incandescent lamps. Because the ballast is connected to the power line, it's the part of the system that must be matched to the proper voltage. The two most common U.S. operating voltages are 120 and 277 V. Once the ballast is connected to its rated voltage, it automatically supplies the lamps with the proper amount of power.

Ballasts consume power regardless of whether the tube is good, burned out, or even removed. "Delamping" usually involves removing two lamps from a four-lamp fixture. Removing the tubes alone gets you about 90% of the potential savings. The ballast must also be disconnected to bring the current draw down to zero.

A word about delamping. This was (and still is) a popular, low-cost way to save energy. While it can work well in halls and other public areas, it can be a disaster where reduced lighting levels have a direct impact on work productivity. Taking two lamps out of a four-lamp fixture not only reduces lighting level, but also creates unequal lighting distribution.

Figure 9-2 shows the relative costs of a typical office on a per-square-foot basis. While reducing lighting levels may save a lot of money, a less-productive workforce can cost you more than the energy savings. Ideally, lighting changes should be unnoticeable to the occupants.

At some time or other, you probably have tried to turn on a small fluorescent light by holding down the switch to make the lamp turn on. Holding down the switch heats two cathodes; releasing it causes a high-voltage pulse to travel across the lamp, causing an arc to strike in the tube. This is a **manual-start system**, Figure 9-3.

Automatic starters were developed that can do the same thing as the manual switch.

Have you noticed fluorescent tubes with only one pin at each end? These are **instant-start systems**, Figure 9-4. An instant-start ballast supplies enough initial voltage so the arc strikes without preheating the cathodes. The base pin acts as a safety switch, eliminating the possibility of electric shock. When the lamp is removed, the circuit is broken.

Rapid-start lamps have the most popular ballast system, Figure 9-5. Separate windings in the ballast provide continuous heating voltage for the lamp cathodes. Under normal conditions, the rapid-start ballast starts the lamps in less than 1 sec. To ensure dependable starting, make sure the lamps are mounted close to an electrically grounded metal strip. (In most cases, the reflector serves this purpose.)

Power Technology

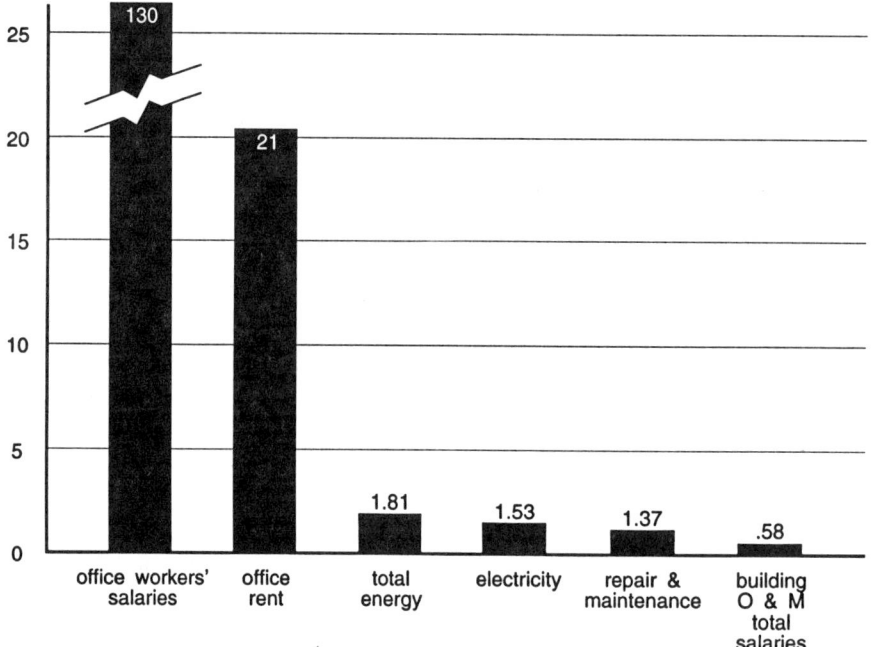

Figure 9-2. Relative cost of a typical office (per square foot)

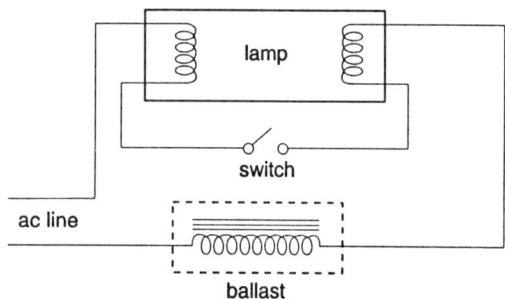

Figure 9-3. Manual-start system

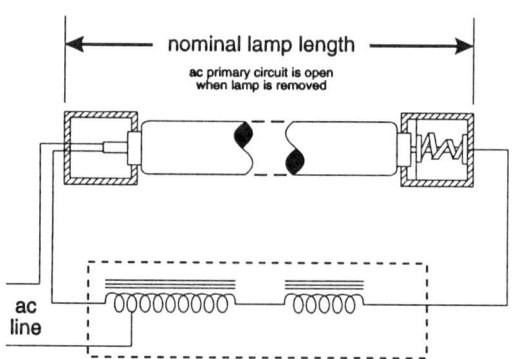

Figure 9-4. Instant-start system

Lighting Terminology, Sources, and Ballasts

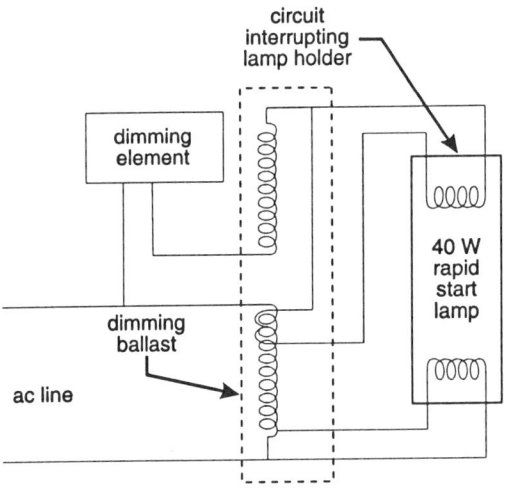

Figure 9-5. Rapid-start lamp with dimming ballast

Two-lamp systems start the lamps in sequence, then switch to series operation. When one lamp is removed from a fixture, the other lamp goes out.

Rapid-start lamps can be dimmed when operating on special dimming ballasts. This ballast keeps the lamp cathodes supplied with the proper heating current, regardless of the lamp's output. Unlike incandescents, the fluorescent lamp's life is not extended (nor shortened) when operating at less than full output.

A main part of a typical ballast is made up of a coil of wire wound around a core of laminated iron. Electronic ballasts use different technology. The circuits change the 60-Hz power line frequency to a higher frequency. Fluorescents are more efficient at higher frequencies; thus, lighting systems that use electronic ballasts are more efficient.

A WORD ON GROUP RELAMPING

Fluorescent lamps can be replaced either individually as they burn out, or in an entire area when the lamps have reached between 60% to 80% of their average life. Although it doesn't seem logical to replace good lamps, it's a cost-effective method of maintaining light levels. The savings realized by the greatly increased labor productivity of group relamping more than offsets any increased material costs.

Regardless of the method used, the luminaires should be cleaned every time a lamp is changed.

Another good reason to turn lights off, especially in air conditioned areas, is that cooling costs will be reduced. As shown in Figure 9-6, fluorescents contribute quit a bit to the heat load. For every dollar of lighting costs saved, up to an additional 40¢ can be subtracted from the cooling bill.

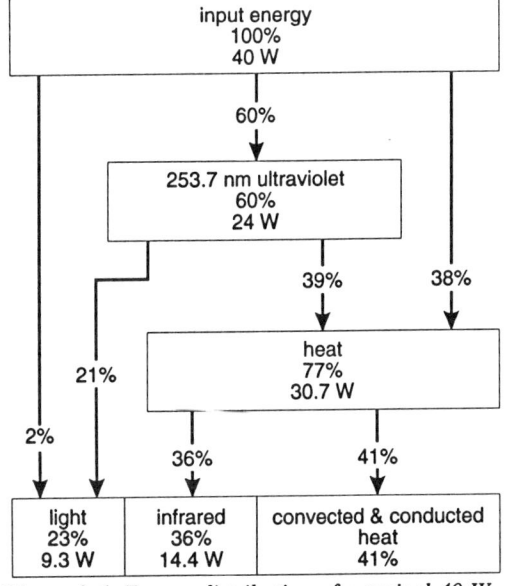

Figure 9-6. Energy distribution of a typical 40-W cool white fluorescent lamp

Chapter 10

MOTORS: THREE-PHASE, SINGLE-PHASE, AND UNIVERSAL

There are probably hundreds of motors in your facility of all different shapes and sizes — from large motors that drive compressors and fans, to small ones found in typewriters and computers.

They can be divided into two general classes: single and three phase. This chapter also covers universal motors, plus cycling, performance, starters, and variable-speed drives.

THREE-PHASE MOTORS: INDUCTION VS SYNCHRONOUS

An engineering evaluation is required when matching a motor to a load, but a rule of thumb is that synchronous motors are preferred when the rating exceeds 1 hp/rpm. The decision to use a synchronous or induction motor depends mainly on size and speed.

INDUCTION MOTORS

Three-phase induction motors are used on just about all large loads because they are efficient and simple to build. Every building with three-phase power has three-phase induction motors.

These workhorses drive every fan and air conditioning compressor with greater than 2 hp. There are only two major parts to three-phase motors: a stator and a rotor.

The **stator**, or frame, is the non-rotating part that holds the windings in place. Attached to both ends of the stator are the end bells, whose main duty is to hold the shaft bearings in place.

The **rotor** is the rotating part that turns the shaft. In most cases, it also turns a fan that cools the motor.

Rotors in induction motors are sometimes called squirrel-cage rotors, because they vaguely resemble a small animal's exercise wheel. Figure 10-1 shows the skeleton of a squirrel-cage winding. Add the shaft and you get the complete rotor assembly.

With induction motors there is no direct electrical connection between the rotor and stator. When a three-phase power source is connected to the stator windings of an induction motor, a rotating magnetic field is produced. This rotating field cuts across the conducting bars in the rotor, causing current to flow. The current then produces a magnetic field in the rotor. The resulting interaction of these two magnetic fields produces torque that causes rotation.

Power Technology

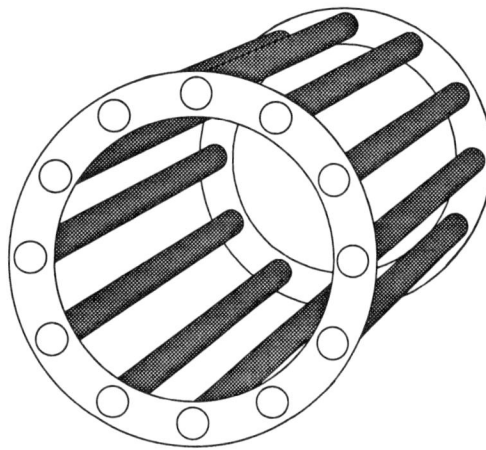

Figure 10-1. Squirrel-cage winding

To better understand how a rotating magnetic field is produced in the stator, look at Figure 10-2 and Table 10-1. As the process proceeds from time 1 through time 6, each phase current rises and falls. Thus, polarity in each coil periodically changes, while the magnetic strength in each coil constantly changes. These continuously changing vectors add-up algebraically to generate a rotating field. (All algebraically means is that you need to pay attention to the sign before the number. For example, the difference between a +6 and a -2 equals 8.)

Reversing the rotation of a three-phase induction motor is easy; just reverse any two of the three power leads. This can be good (it's easy to do on purpose) or bad (it's also easy to do by mistake), depending on what you meant to do. When a centrifugal pump or fan operates backwards, the mistake isn't readily apparent until the unit reaches about half capacity. Then, although all other points of operation seem normal, the remainder of the unit's expected performance just isn't there.

If the stator is wound and connected to a single set of north-south poles, it would drag the rotor around at 3,600 rpm (two poles = 3,600 rpm). If the stator was set

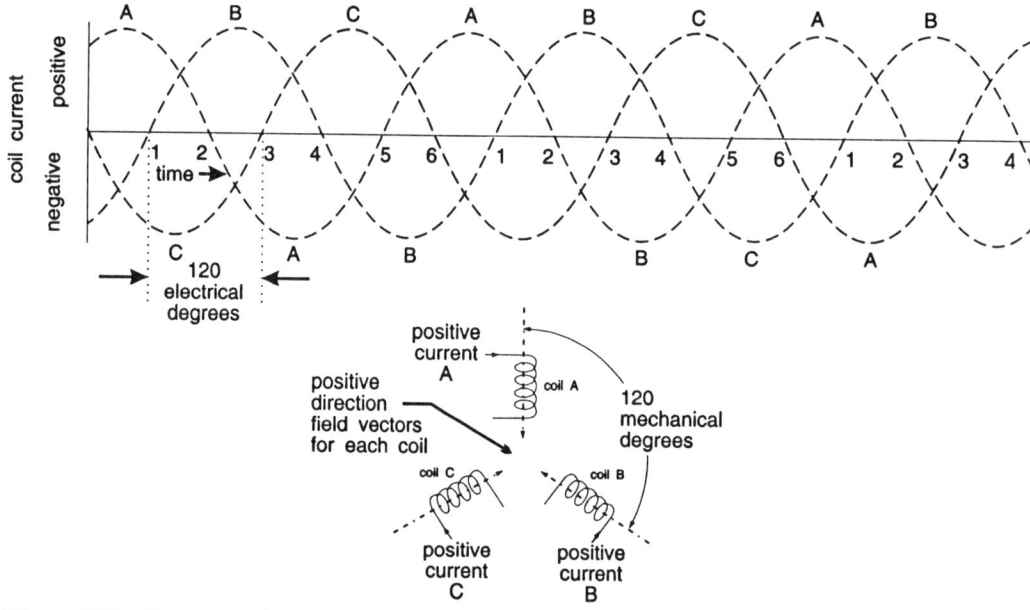

Figure 10-2. How a rotating magnetic field is produced in the stator

Time	Coil number	Percent of maximum coil current	Coil current direction	Direction of coil magnetic-field vectors	Vector addition and degrees total magnetic vectors rotation*
1	A B C	86% 0 86%	positive negative	same as coil A opposite to coil C	ϕ_1 with C, A vectors
2	A B C	0 86% 86%	 positive negative	 same as coil B opposite to coil C	ϕ_2 with B, C vectors
3	A B C	86% 86% 0	negative positive 	opposite to coil A same as coil B 	ϕ_3 with A, B vectors
4	A B C	86% 0 86%	negative positive	opposite to coil A same as coil C	ϕ_4 with A, C vectors
5	A B C	0 86% 86%	 negative positive	 opposite to coil B same as coil C	ϕ_5 with C, B vectors
6	A B C	86% 86% 0	positive negative 	same as coil A opposite to coil B 	ϕ_6 with B, A vectors
and repeat		* note clockwise rotation of equal magnetic vectors ϕ top to bottom t total vector ϕ rotation between time 6 and time t is 60 degrees			

Each step shows 60° rotation.

Table 10-1. Magnetic field vectors determined for one cycle; per coil and total (ϕ)

up with two or more sets of poles, the rotor would spin at a slower speed. (For example, four poles = 1,800 rpm and six poles = 1,200 rpm.) The equation for the synchronous motor speed is:

$$\text{rpm} = \frac{\text{Hz} \times 120}{\text{poles}}$$

Notice I said synchronous speed. In order for the rotating stator field to produce a magnetic field in the rotor, the rotor must rotate at a slightly slower speed. This "slip" is very small at no load, and increases to a few percent at full load. For example, a four-pole induction motor actually turns at 1,750 rpm. The lost 50 rpm is the slip. This slip is necessary for the motor to work.

SYNCHRONOUS MOTORS

Synchronous motors can easily operate at low rpm and are therefore excellent for driving low-speed refrigeration and air compressors. If a compressor requires 514 rpm, the motor can be coupled directly to the machine. This makes them competitive with induction motors, which require speed reducers for this type of application. Synchronous motors can also go as high as 3,600 rpm. In short, a large, high-speed synchronous motor is more efficient than an induction motor.

Synchronous motors usually aren't found in commercial or light industrial situations, except for plug-in electric clocks, which use a type of synchronous motor.

Unlike an induction motor, a synchronous motor follows the rotating stator field exactly, because the rotor is supplied with magnetizing power and thus no slip is required to produce the rotor's magnetic field. Small synchronous motors use a permanent magnet, larger ones use an electric magnet. Electric magnets require a source of dc power called **excitation**. Excitation has traditionally been supplied to the rotor via slip rings and brushes, but lately "brushless" machines have become common.

When starting, the synchronous motor accelerates like an induction motor by using the squirrel-cage windings on its rotor. Meanwhile, a resistor placed across the rotor's electric magnet prevents destructive high voltage from building up. Near full speed, the resistor is disconnected and the magnet is energized to pull the rotor into synchronization. Figure 10-3 shows how a motor's speed and hp operate for both synchronous and induction motors.

Synchronous motors can be more expensive and require more maintenance than induction motors, but remember their chief advantage: They are more efficient, especially at lower speeds. An efficient, high-horsepower, constantly running motor can quickly pay back the premium originally paid for it. Also, their power factor can be regulated by adjusting rotor excitation. (See Chapter 5 on power factor.)

SINGLE-PHASE MOTORS

Single-phase motors are small, usually under one horsepower, and they come in a variety of designs.

Operating a properly designed motor running on single phase is no problem. However, *starting* a motor on single phase from complete rest requires help from a starting coil, to help bring the unit up to speed. The main and starting windings are designed so that when both are energized, the motor thinks two phases are being supplied, thus developing the required starting torque.

Once up to speed, this auxiliary or starting winding needs to be removed from the circuit. This is usually accomplished by a centrifugal switch. At rest, this switch is closed and the starting winding is in the circuit. When the motor gets up to speed the switch opens, removing the start winding, and only the main or run winding remains energized. (If you've ever heard a click as a single-phase motor slows down after it has been turned off, this is the centrifugal switch closing.) Because of this extra starting gear, single-phase motors are more expensive than three-phase motors for a given horsepower.

Most single-phase motor-starting schemes are either the split-phase or capacitor types.

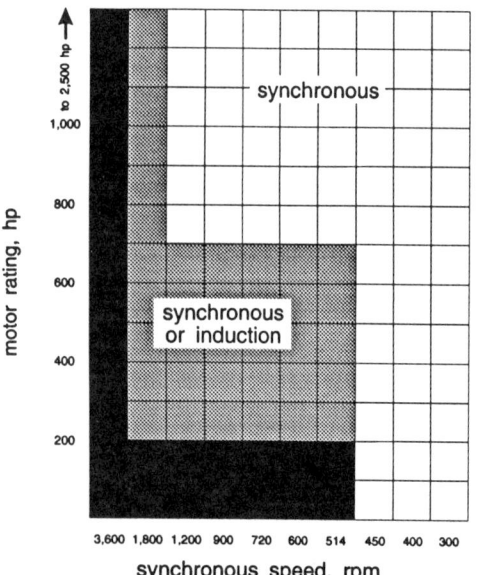

Figure 10-3. Motor rpm and hp for synchronous and induction motors (black area—induction motors only)

SPLIT-PHASE STARTERS

When low starting torque is all that you require, an inexpensive means to achieve this is to form the starting winding with smaller wire. This gives it more resistance, which puts it out of phase with the run winding. Split-phase (or resistance split-phase as it is sometimes called) is inexpensive to make, and develops enough starting torque for light loads, such as small fans.

CAPACITOR STARTERS

Capacitor-start motors are more expensive, but they can start high-inertia loads just as well as three-phase motors. A capacitor is placed in series with the starting winding. The starting winding is made of heavier wire, so it has the ability to handle more current. Because capacitors do a good job of phase shifting, and the starting winding is able to carry more current, more starting torque is produced.

In both cases, if the starting windings are not taken out of the circuit as the motor comes up to speed, they will be damaged.

UNIVERSAL MOTORS

Universal or series motors can operate as high as 35,000 rpm. They are used on handheld drills, routers, blenders, etc., and are called "universal" because they operate on either ac or dc. The rotors have wire-wound coils; they are also called "series" because the stator and rotor windings are all wired in series. The electrical connections between the stationary stator and the rotating rotor are made with carbon brushes.

ELECTRIC MOTOR CYCLING

Cycling a motor on and off for specific intervals is often done to control processes, or as an energy management strategy. Caution must be exercised, to make sure motors are not damaged from repeated starts over a short time interval.

Figure 10-4 shows the thermal effects of cycling a motor. When a motor is turned off there is a slight rise in temperature followed by a slow cooling period. On restart, due to the high starting inrush current, there is a rapid temperature rise followed by a gradual cooling down to normal operating temperature. If the off time is too short, the motor doesn't have a chance to cool down. The net result is an overheated motor and an eventual burnout.

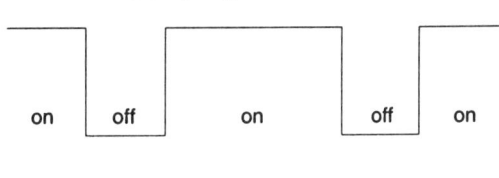

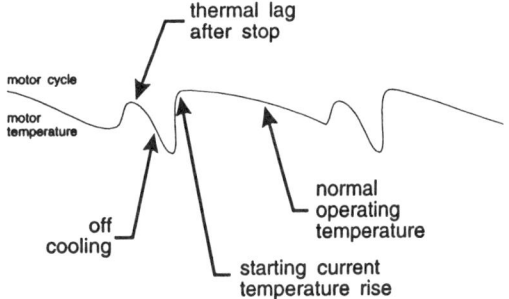

Figure 10-4. Thermal effects of cycling a motor

Table 10-2 shows guidelines developed by the National Electrical Manufacturers Association (NEMA). For some small motors, 3 min of off time is all that is required. For motors larger than 50 hp, an engineering evaluation should be made.

Motor hp	¼ - 10	10 - 20	20 - 50	50 - 100*
1. Minimum interval between starts (on time + off time minutes)	10	20	30	40
2. Minimum off time (minutes)	3	5	7	7
3. Suggested maximum off time (minutes to maintain comfort)	5	7	10	10
4. Minimum % kW savings	30%	25%	23%	17%
5. Reciprocating compressors **Min. interval	50	50	50	60
5. Reciprocating compressors ***Min. off	15	15	15	20

* Observe several start/stops for unusual noise, belt slippage, etc.
** See item 1 of this table.
*** See item 2 of this table.

Table 10-2. NEMA motor-cycling guidelines

The energy savings from cycling are great, but it is advisable to contact the motor's manufacturer for advice.

VOLTAGE, MOTOR PERFORMANCE, AND PREMATURE FAILURE

It is extremely important to operate a motor at its rated voltage. Manufacturers go to great lengths to have their motors operate as advertised at the specified nameplate voltage. Figure 10-5 shows the effects of improperly applied voltage. Notice the increase of slip and full-load current at lower voltage. If a motor draws excessive current over a period of time, it could soon experience overheating and premature failure.

One of the most common occurrences of voltage misapplication is with 208- and 240-V motors; 208 V is common in commercial and industrial situations, while 240 V is common in residential situations. An air compressor or air conditioner is purchased with a 240-V motor, then connected

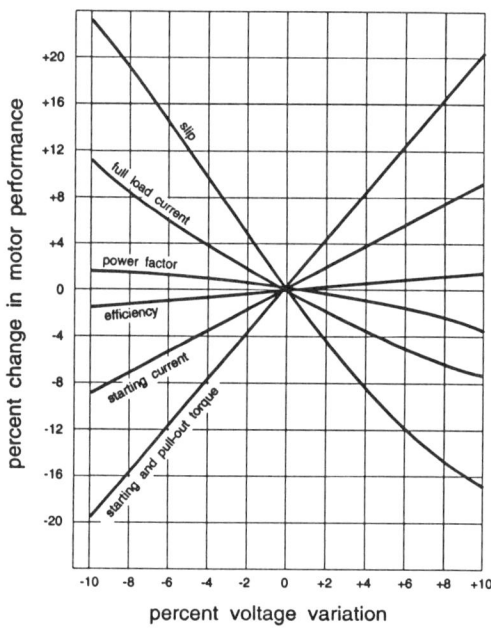

Figure 10-5. What happens to motor performance when voltage varies

to a 208-V source. Sometimes you can get away with it, sometimes you can't. This mismatch can easily be corrected with autotransformers.

Another factor leading to premature failure of polyphase motors is unbalanced voltage. A relatively small voltage imbalance causes a considerable increase in temperature rise. In the phase with the highest current, the percentage increase in temperature rise is approximately two times the square of the percentage voltage imbalance. To illustrate, a 3.5% voltage imbalance causes a 25% increase in temperature rise. (Ideally, the imbalance should not exceed 1%.)

STARTERS

When motors start, they draw up to six times their normal full-load run current. Heavy-duty switches or contacts are required for starting motors. In conjunction with these contacts, additional devices protect the motor from excessive current when operating. In the case of combination starters, fuses or circuit breakers protect the feeder wires from overcurrent.

Pictured in Figure 10-6 are manual starters for small motors. They look like wall switches, but inside these units are heavy-duty switch contacts and a device that protects motors from overloads. These devices are called **heaters**, because they are sensitive to the heat produced by the motor's current which is passing through them. (These heaters are schematically represented by two backwards-joined question marks.)

Heaters must be matched to the motors they protect. Manufacturers provide extensive tables to aid in their selection. When a load is switched, all ungrounded or hot conductors must be opened. In Figure 10-6, the starter on the right has two sets of contacts

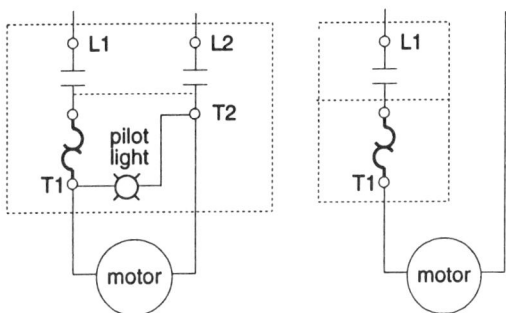

Figure 10-6. Manual starters for small motors

to accomplish this. (Dotted lines on an electrical drawing denote a mechanical connection.)

Figure 10-7 shows a schematic for a typical three-phase power, across-the-line motor starter. Instead of the contacts being closed manually, they are pulled in with a magnetic coil (schematically represented by a circled M). It used to be common practice for this coil and control circuit to be powered directly from the line voltage. Now, to make the equipment safer, starter control circuits operate at 120 V, which is often supplied from a built-in step-down control transformer.

When the start button is pushed, the contact coil is energized and the three-phase motor contactors close. (Contactors are shown as two parallel lines.) A smaller set of auxiliary contacts connected in parallel with the start button also close. These **sealing contacts** remain closed after the start button is released, supplying power to the main motor contactor coil. The contactor drops out when:

1. The stop button is pushed.

2. A motor overload opens.

3. Power is lost to the starter.

Power Technology

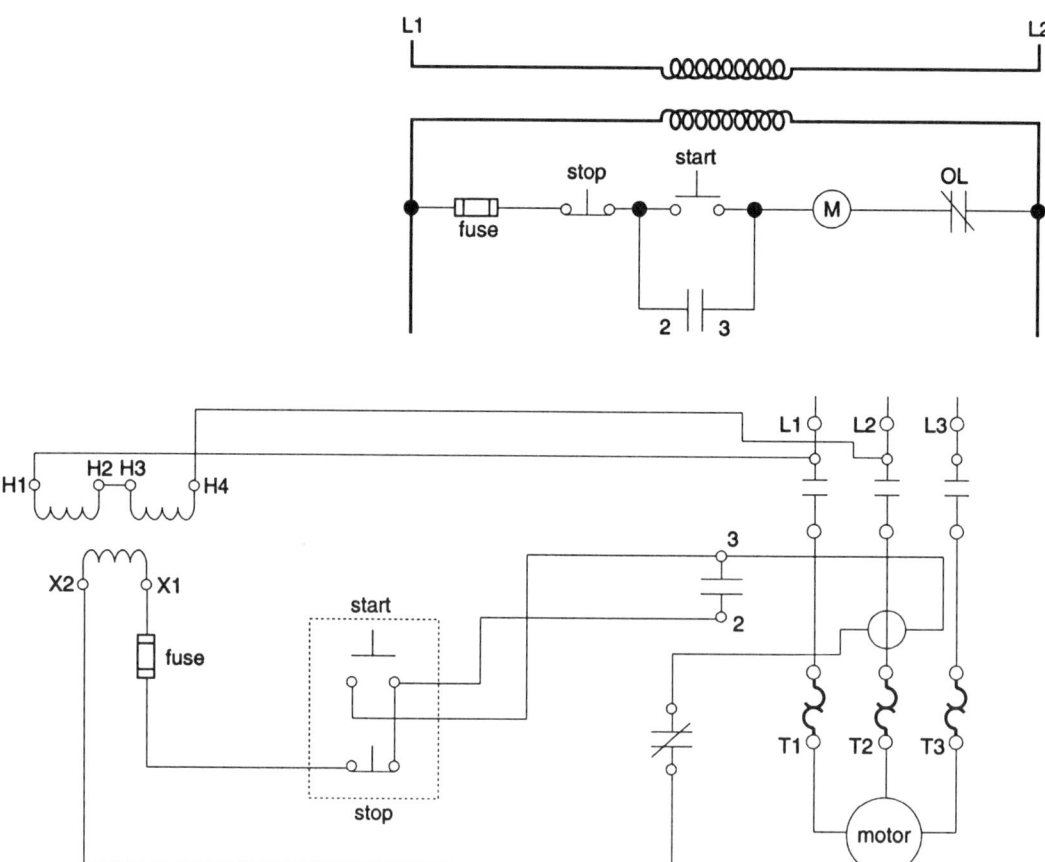

Figure 10-7. Schematic for typical three-phase power, across-the-line motor starter

Starter contacts are heavy-duty devices that can handle the inrush of current required to start a motor from rest. Typical inrush is six times the motor's normal amperage.

Ever notice how lights may dim when motors start? This voltage dip is caused by the current inrush and is a symptom of inadequate electrical capacity. A system that allows large motor starting without excessive voltage dips is referred to as a "stiff system."

SPECIAL STARTERS

When the load allows a motor to reach top speed rapidly, it can be started by a direct connection to the line — across the line starting. If, however, a high-inertia load results in a long acceleration time, then the starting method needs to be changed. Hydraulic elevator motors, conveyer belts, and loaded compressors are examples of such high-intertia loads.

Of the many methods for starting high-inertia loads, the wye-delta method has been quite popular. When a motor is wye connected, the winding voltage is reduced to 58% of its normal voltage. This limits the motor current to 33% of normal locked-rotor line current and thus reduces stress while the motor is accelerating. When up to speed, the starter reconnects the motor to normal delta operation. Figure 10-8 shows a schematic of a wye-delta starter, along with a list of other starting methods.

SOLID-STATE STARTERS

Recent developments allow solid-state technology to connect power to motors. Conventional contacts are either opened or closed, while solid-state contacts can be programmed to limit voltage and current inrush. If a load takes a long time to get up to speed, solid-state starters can limit starting current to a predetermined level and prevent excessive stress to the motor.

Solid-state starters can also "soft start" and "soft stop." With an across-the-line starter, connected loads are mechanically shocked when coming up to speed. With a soft start, the load is gradually accelerated with a smooth, bump-free transition up to full speed. This adds life to drive belts and other transmission components.

When some loads are brought to a rapid halt, operational problems can result. An example is shutting off a large pump abruptly and having the inertia of the water cause water hammer damage. With an orderly shut-down programmed into a solid-state starter, such problems are avoided.

Solid-state technology has made it possible for even small motors to have sophisticated protection. Microprocessors coordinate inputs such as under and unbalanced voltage, ground fault, excessive and unbalanced current, time between starts, and they make more informed decisions regarding when to trip a motor.

VARIABLE-SPEED DRIVES

To be absolutely, technically correct, we should call them variable-frequency drives. But, if you hear them referred to as variable-speed drives, don't panic; they're one and the same.

Used in industry for many years for precise control of machinery like printing presses, they now find a wide application to save energy for pumps and fans.

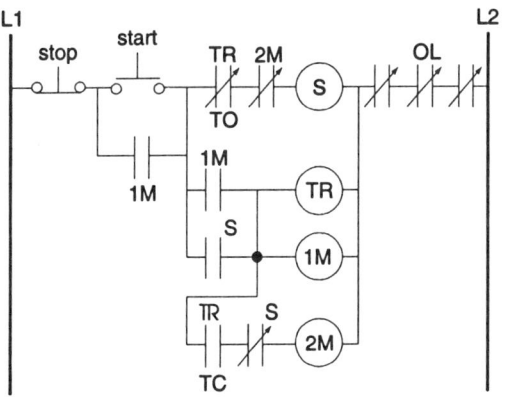
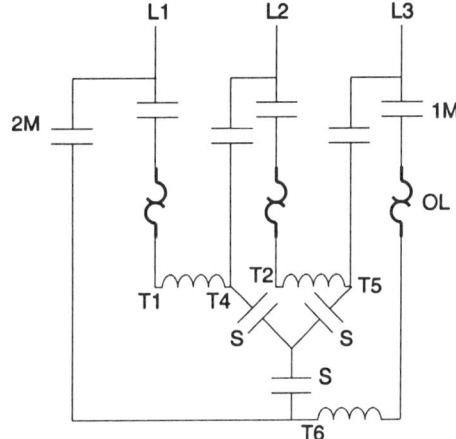

Figure 10-8. Schematic of a wye-delta starter

Fan and pump systems are designed to meet peak demand, with perhaps some extra thrown in to cover unexpected situations. The problem is, this peak demand occurs for only a few hours a year. The rest of the time the system is oversized.

It makes sense that slowing down a fan or pump saves energy, but the savings are more than you think. I am going to throw some equations at you, just to show where the following statements come from.

Equation 1:

$$\frac{rpm_1}{rpm_2} = \frac{flow_1}{flow_2}$$

Equation 2:

$$\left(\frac{(rpm_1)}{(rpm_2)}\right)^2 = \frac{pressure_1}{pressure_2}$$

Equation 3:

$$\left(\frac{(rpm_1)}{(rpm_2)}\right)^3 = \frac{hp_1}{hp_2}$$

Zero-in on the rpm/hp relationship, which states that horsepower is proportional to the cube of the rpm. Translated into English, a small reduction in rpm means a large savings in energy. For example, if speed were reduced 20%, the theoretical power required would be reduced 48%. Studies have shown that a lot of fan and pump systems operate at 70% to 80% of capacity. At 70% capacity, the theoretical power required drops to 35%. (Notice I said "theoretical" power; the variable-speed drive needs some power to operate. Check with the vendor to find out how much.)

If your evaluation shows that a system runs at or near capacity, then a variable-speed drive is not justified on the basis of cost savings.

Fan dampers and pump throttle valves regulate flow and reduce power, but variable-speed drives do it more efficiently, Figure 10-9.

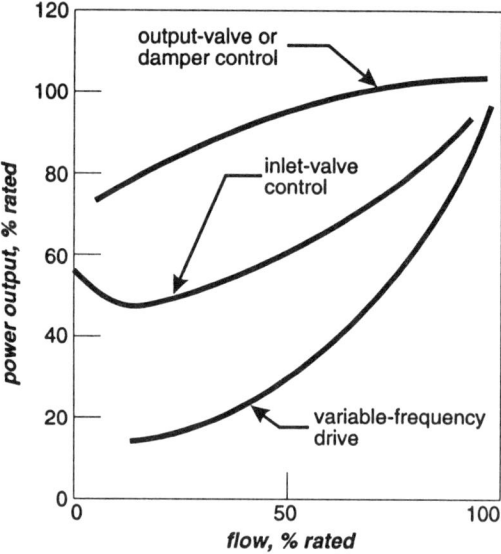

Figure 10-9. Efficiency comparison of fan dampers, variable-speed drives, and variable-speed drives for regulating flow

The great majority of off-the-shelf induction motors have no problem operating with a variable-speed drive. You should check with the motor's manufacturer to make sure.

Another advantage is the sophisticated motor protection that is standard with new variable-speed drives. When the microprocessor sees any abnormal situation, it shuts down the drive and a display tells you what happened.

MAINTENANCE

As with anything that produces, transmits, or converts energy, electrical systems require maintenance. This chapter covers infrared diagnostics; testing and maintaining oil-filled and dry-type transformers; PCBs; motors; cleaning; and circuit breakers.

INFRARED SCANNING

All electrical devices produce a specific amount of heat. When a device generates more heat than expected, this usually indicates some sort of trouble.

A good way to determine the condition of an energized distribution system is to scan it with an instrument that uses infrared light. Infrared scanners read surface temperatures without having to make physical contact.

Infrared testing is usually a two-person job. An electrician removes panel covers, or otherwise provides a line of sight to the equipment to be tested. The other person then points the scanner at the area in question and reads its temperature. With infrared scanners, hot spots show up immediately.

Most scanners can take pictures; a photographic record can aid the repair of problem areas if kept in maintenance files. And, if infrared scanners are used in routine maintenance, many minor problems (such as loose connections) can be detected before they turn into big-time trouble.

Some in-house maintenance organizations keep this equipment on hand, but it is more common to bring in a contractor for the job. There are two main advantages: You don't need to purchase and maintain expensive equipment, and experienced contractors generally interpret the data better.

OIL-FILLED TRANSFORMERS: TESTING AND PCBs

The oil in oil-filled transformers insulates and carries heat away from the windings. This design allows a great deal of power to be transformed at high voltage, in a relatively small package. Oil-filled transformers are also less noisy than air-cooled transformers.

Mineral and PCB (polychlorinated biphenals) are the two most common oils used in transformers. Because of the fire potential, oil-filled transformers must be installed outdoors, or in a vault if installed indoors. PCB-based oil used to be quite popular because of its high flame resistance and excellent insulating properties. However, PCBs are no longer manufactured. We will discuss this later in the chapter.

OIL TESTING

Annual transformer oil testing helps determine the internal condition of liquid-cooled transformers, much the way a series of blood tests helps determine a person's physical condition. The main tests performed on oil analyze its:

- Color
- Insulation value
- Moisture content
- Acidity
- Dissolved gases

Color: Under normal conditions, the oil's color remains the same throughout its service life. Although somewhat subjective, any change indicates an abnormal condition, and gives the technician a quick indicator of the oil's general condition. By no means should the color test be used alone. The sample for mineral oil is taken from the bottom, so sludge and other heavy matter has a good probability of showing up in the sample. Water would not because most of it would be on or near the surface. The sample for PCB oil is taken from the top, so the probability of sludge in the sample would be less.

Insulation value: If the insulation or dielectric value of the oil is low, internal shorts between windings are more likely. Moisture is one of the factors that lowers oil's insulation value. Usually, the oil's quality can be improved on site by filtering. The transformer must be shut down for this process, and left off for up to 24 hours after filtering to get rid of all the air bubbles.

Moisture content and acidity: Moisture also combines with other contaminants to form acidic sludge. Measuring the oil's pH level is an easy way to detect this condition. Sometimes filtering can take care of this problem too. If severe, however, the oil must be changed. (In extreme cases, the transformer must be taken to a shop for dismantling and cleaning.)

Dissolved gases: Dissolved gas analyses is a relativity new procedure. One series of tests looks for fuel gases in the oil. Correcting the conditions that introduced the gas into the oil can save the transformer's life.

For example, the presence of carbon monoxide in oil indicates that mild arcing has taken place; probably no immediate action is required, but you'd want to closely monitor future tests. Acetylene indicates that severe arcing has taken place. Immediate action is required; the transformer is on its way to failure.

PCBs

PCB-filled transformers require the most maintenance. Although a near-perfect, fire-resistant cooling medium, when PCBs do burn, their products of combustion are extremely dangerous. For this and other reasons, PCB transformers are no longer manufactured, and the existing stock is highly regulated. In fact, in locations such as food-processing and health-care facilities, it is illegal to have PCB equipment on the premises.

Even where allowed, there are a large number of regulations on their operation, labeling, testing, and disposal. There must be enough signs displayed so everyone knows where the PCBs are. Records must to be maintained on all aspects of the equipment's operation, inspection, testing, and maintenance history. Many local Fire Departments require to know the PCBs whereabouts in a facility, and have input on the company's detailed emergency evacuation plan.

When the time comes to get rid of the PCBs, there are two choices: retrofit or replace. Each case must be evaluated on its own merit. To make an informed decision, ask the following questions:

- Is the equipment reaching the end of its useful life?
- Is the equipment of special design that can't easily be replaced?
- Is the equipment so inaccessible, just the rigging to remove and replace it will cost big bucks?

Retrofitting equipment to use non-PCB oils can be an economical decision in some circumstances. The retrofit process usually costs about 70% to 80% of what a new replacement costs.

After the transformer is taken out of service, the first step in retrofitting is to drain it.

Note: *Make sure the contractor is well versed in regulations regarding the disposal of PCBs and other hazardous wastes. Government fines for violating hazardous waste disposal regulations can put your company out of business; bad publicity for toxic dumping can drive the nails into the coffin.*

Next, the transformer is flushed a few times before being returned to service. (All the flushing fluid also must be properly disposed of, since it is PCB contaminated.)

Now PCBs must be cleaned out of the nooks and crannies between the windings and in paper insulators used to separate various internal parts. Given time, these PCBs would leach out and contaminate the fresh oil. The draining-flushing operation must be repeated at least one more time. The transformer is still considered contaminated if the oil contains more than 50 ppm (parts per million) of PCBs.

Replacement also has its pitfalls. Say the new apparatus is in place and the old equipment is on its way to the landfill. The problem is, *that old transformer is yours forever.* Even if all regulations and procedures were followed to the letter of the law, if something beyond your control goes wrong, *you are libel.*

This is why some organizations take an extra step: They have the transformer disassembled and sold as scrap. You still have the liability, but not the exposure, since the carcass is now part of a car or steel beam, the copper turned into house wiring, and the aluminum into beer cans.

Regardless of what option is used, it's in the company's best interest to go with the most-effective method, not the cheapest. Check on the contractor's reputation; try to determine if they will be there down the road.

DRY-TYPE TRANSFORMERS

Dry-type transformers, which are cooled by air that circulates through their windings, are relatively maintenance free. Just keep them clean so air can circulate freely, and perform an infrared scan to look for loose connections.

Noise can create problems for air-cooled transformers. However, they are easily solved when a technician removes the covers and tightens the transformer's component fasteners back to their original specifications.

MOTOR TESTING

The most common way to test motors has been to use a megger (megohmmeter) and "hipot" to test resistance to ground. A megger can read high resistance values, up to several million ohms. Most models have hand-cranked generators to produce the voltage necessary for the test.

A hipot (short for high potential) has a variable dc voltage supply. As voltage is increased, the amount of ground leakage can be read in microamps. When the current reaches a predetermined level, the voltage producing this current gives an indication of the insulation's condition.

These two conventional tests aren't 100% reliable for predicting motor life. Newer methods and microprocessor-based equipment have been developed to more accurately forecast motor mortality.

Note: *All these tests must be performed with the equipment shut down and disconnected from its power sources.*

Motors fail mechanically as well as electrically. Bearing failure accounts for a large percentage of motor down time. So, even when a failed motor is on the bench undergoing an "autopsy," it's sometimes hard to determine the cause of death.

The best way to minimize mechanical failure is by proper lubrication. Also, it cannot be overemphasized that the manufacturer's recommendations be followed to the letter.

LUBRICATION AND BEARINGS

When in operation, bearing surfaces are continuously separated by a film of oil so there is no metal contact. The bearing faces of the shoes take on an inclined position, so the oil film between the shoes and collar are wedge shaped (the thin end points in the direction of rotation).

Maintaining well-lubricated bearings is an important part of any equipment maintenance program. This section describes different types of bearings and the application of grease to the anti-friction type.

BEARINGS

The function of a bearing is to support the weight and control the motion of a rotating shaft while consuming a minimum amount of power. There are two main types of bearings: **anti-friction** and **sleeve**.

Anti-friction bearings use balls or rollers to convert sliding friction into rolling friction, Figure 11-1. They are constructed by placing these balls or rollers between an outer race and an inner race. A retaining cage or separator keeps the rolling elements equally spaced. Some bearings also have shields to keep lubricants in and dirt out.

Sleeve bearings depend on a film of oil to support the shaft so there is no metal-to-metal contact. The oil is dragged along by the rotation of the shaft and forms a wedge between the shaft and bearing, Figure 11-2.

Sleeve bearings are found on small appliances, such as fans, because they are cheap. They are also used on large pumps, turbines, and engines where anti-friction bearings are inadequate and/or impractical. These larger bearings are made of a babbitt lining, $1/8$ in. thick or more, and are anchored in a cast-iron or bronze shell. (Babbitt is an alloy of tin, antimony, lead,

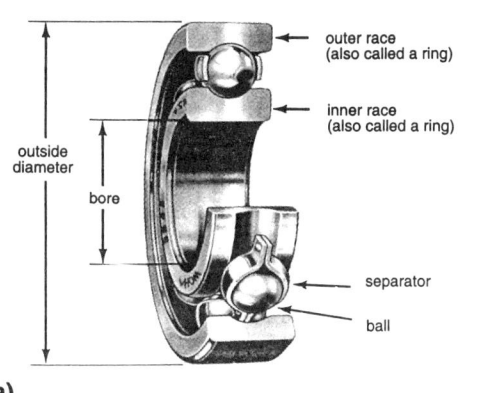 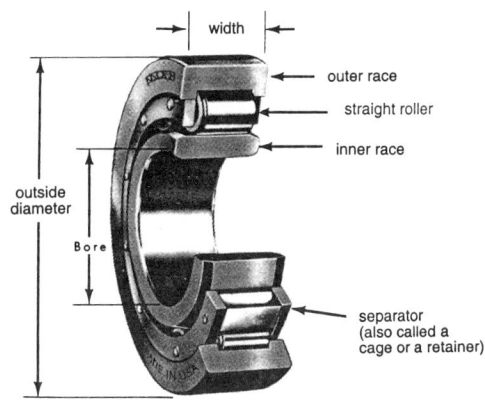

Figure 11-1. Anti-friction bearings: a) ball type, b) straight roller type

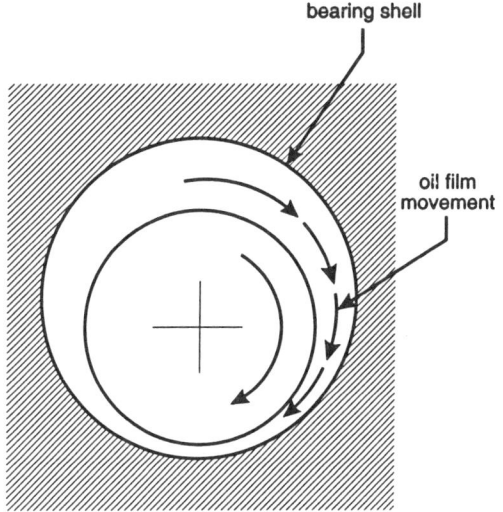

Figure 11-2. Sleeve bearing

and copper, with the ability to "wipe" away at a small spot where metal-to-metal contact exists, until an oil film is established.)

For heavy-thrust loads, **Kingsbury thrust bearings** are commonly used. The axial load is transmitted to the stationary bearing shoes through a forged-steel collar rigidly attached to the shaft. Leveling plates equalize the load among the shoes, which are free to tilt both radially and tangentially as required to compensate for operating conditions.

The type of lubrication used on bearings depends on application requirements. Refinery engineers prefer oil, while marine engineers prefer grease. Generally, oil is preferred for high speeds (in excess of 5,000 rpm).

GREASE

Grease packed into a bearing is thrown out by the rotation of the balls, creating a slight suction at the inner race. As heat is generated in the bearing, the flow of grease accelerates until it is thrown out at the outer race. The expelled grease cools on contact with the housing and is attracted back to the inner race. This continuous circulation of grease lubricates and cools the bearing.

As a lubricant, grease is quite effective and requires little attention. It is preferred on vertical shaft bearings because there is less chance of leakage.

A fully packed bearing housing prevents proper grease circulation. Well-established practice recommends that only one-third of the housing be filled. An excess amount causes overheating unless the grease can flow out of a seal or vent. If this built-up pressure is not relieved, the bearing eventually fails.

How to apply motor-lubrication grease on anti-friction bearings:

When equipped with fitting and drain —

1. Remove the drain plug at the bottom of the bearing housing; clean out any hard grease.

2. Wipe the grease fitting clean.

3. Add new grease through the fitting until the old grease has been purged through the drain and new grease begins to appear. If possible, add grease with the motor running.

4. With the drain plug removed, allow the motor to run at operating temperature. This allows the grease to expand so the excess is forced out of the drain. Excess grease stops draining when normal pressure is reached in the bearing, usually within 10 to 30 min.

5. Clean and replace drain plug.

When equipped with fitting but no drain —

1. While the motor is running at operating temperature, remove the fitting to purge the excess grease.

2. Clean and replace fitting.

3. Pump a small amount of grease into the bearing, taking care not to rupture the grease seal.

4. Remove the fitting and allow motor to run for several minutes to purge excess grease. (If no grease comes out, the bearing must have been dry. Repeat steps 3 and 4 until excess grease comes out.)

5. Replace fitting.

Anti-friction bearings can also be lubricated with oil. The proper oil level is at the center of the lower-most ball.

A **constant-level oiler** increases the capacity of the reservoir. It works like an upside-down bottle on a water cooler. As long as the oil level is higher than the inverted bottle opening, atmospheric pressure holds the oil in. When the opening is exposed, the weight of the oil allows it to run out. When the level is high enough to cover the opening again, the oil flow stops. The level is adjustable by moving the inverted bottle up or down.

Ring oiling is popular for smaller applications. A soft steel ring rides on the shaft through a slot cut in the top half of the bearing shell. The ring rotates as the shaft turns, picks up oil from the reservoir in the housing, then wipes it off on top of the shaft. The oil then flows between the shaft and the bearing and discharges through the ends.

Forced lubrication is required in larger machines, especially those with thrust bearings. Usually the system's main pump is driven off the shaft of the machine being lubricated; an auxiliary pump is driven by a small electric motor. The auxiliary pump is used before start-up to circulate the oil and establish oil pressure. The control circuit prevents the large machine from starting unless the oil pressure interlock is satisfied.

When the machine is up to speed, the main lube oil pump is placed in service and the auxiliary lube oil pump is placed on standby. It starts up automatically if the shaft-driven pump fails.

Larger machines can add a considerable amount of heat to the oil. In these cases, oil is cooled by circulating water through coils submerged in the oil reservoir or through a heat exchanger.

CLEANING

Dirt limits heat transfer, and it limits the transmission of electrical current between contact points. When dry, it act as an insulator; when wet, it acts as a conductor and is often the cause of short circuits.

Make an effort to dismantle equipment to the extent where it can be cleaned. In most cases it is necessary to de-energize the equipment first.

For example, after securing the power from air-cooled transformers, the top and sides should be removed and the inside thoroughly vacuumed, and the panel board and starter covers should be removed and checked for dirt.

I once worked in a copper refinery that generated so much dirt, the equipment should have been shut down and cleaned every six months. Production schedules only allowed cleaning once every three years, usually when there was a strike and the office staff was pressed into service. On the other hand, I also worked in a clean research facility where once every three years was probably more than enough, but they tore everything apart and cleaned yearly.

BREAKER TESTING

Circuit breakers can sit around for years without being touched, but when a fault occurs, they had better work as intended. Therefore, periodic testing is required. The bigger the breaker, the more there is to lose if it doesn't work, and the more often it should be tested.

With large breakers, the sensing element is usually separate from the breaker itself. The trip signal from the sensing element can be verified without shutting the system down by disconnecting it from the breaker's tripping mechanism. This proves that everything is in order up to, but not including, the breaker mechanism.

In addition to relay testing, all large breakers should be exercised yearly — for large molded-case breakers, maybe every three to five years. Breakers on small power panels are rarely tested.

Chapter 12

ELECTRIC METERS

Before we discuss electric meters, think about this quotation from William Thomson Lord Kelvin, the person who, among other things, first suggested the use of the gas thermometer for accurate temperature readings and developed the Kelvin Thermodynamic Scale in the late 1800s:

"When you measure what you are speaking about, and express it in numbers, you know something about it; but when you cannot measure it, when you cannot express it in numbers, your knowledge is of a meager and unsatisfactory kind."[1]

Kelvin wasn't thinking specifically about electric meters when he said this, but you get the picture.

Unlike physical media, such as water, electricity is silent and invisible. You can't judge the force (pressure) and amount of electricity by watching it flow out of the end of a conductor. But even though it's invisible, electricity can make its presence known. Have you ever been zapped when changing a switch or outlet? You couldn't see the electricity, but you sure could feel it.

Being able to measure electricity helps prevent such accidents, as well as aids in power system troubleshooting. Instruments are available that convert the characteristics of electricity into its two basic measurements: **voltage** and **amperage**. From these follow other values, such as **resistance** and **power**. The instruments that measure electricity and its offshoots include analog and digital meters, ammeters, voltmeters, ohmmeters, multimeters, meggers, and wattmeters.

ANALOG METERS

For years, most meters applied D'Arsonval-type movements, which consists of a stationary permanent magnet and a movable coil mounted on a shaft, suspended by two jewel bearings. Figure 12-1 shows a typical analog meter. When current flows through the coil, the resulting magnetic field reacts with the permanent magnetic field, causing the movable coil to rotate. The rotation is restrained by springs. The more current going through the coil, the greater the rotating force developed, and the more the shaft rotates. A pointer mounted on the shaft indicates the amount of rotation. Although requiring watch-like tolerance, this movement can be made rugged enough for field use. It is still quite popular.

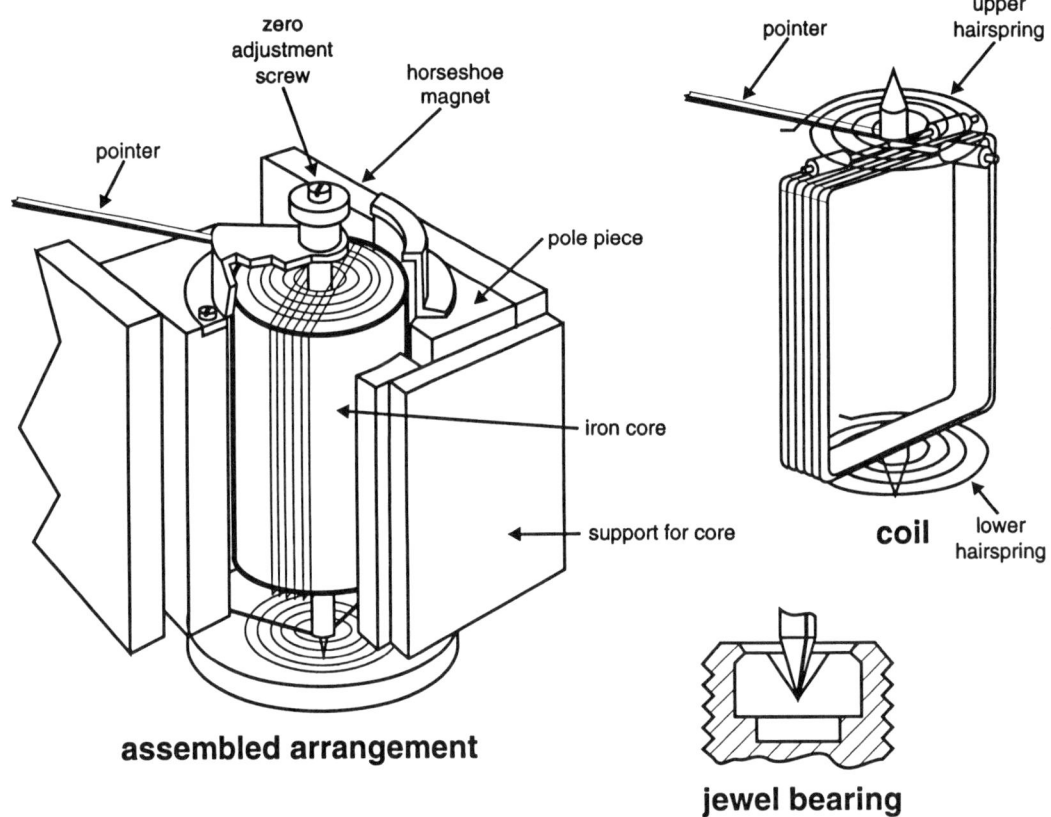

Figure 12-1. Typical analog meter

Taut-band movements are more rugged than the D'Arsonval type. The pivots and jewels are replaced by flat springs that act like torsion bars. Current passing through the movable coil rotates the pointer. The tendency of the stretched, twisted springs to flatten out returns the pointer to zero.

DIGITAL METERS

Digital meters have become quite popular. Notice the ∿ and ⎓ symbols, Figure 12-2. When the rotary switch is turned to a ∿, the meter is taking ac measurements. When turned to a ⎓, it's taking dc measurements.

The next type of metering device to come along read values digitally. Here a signal is reduced to binary form — that is, a series of zeros and ones — processed, then converted back to readable "human" form. Sounds like the long-way around, but the results are more precise. Now, instead of squinting at the graduations on a glass tube, the numbers are displayed in easy-to-read digits. The digital signal can also be fed into a communications network, if desirable.

Look what digital technology has done to the music recording industry. In digital form, playback fidelity has never been better.

Electric Meters

Figure 12-3 shows a typical block diagram of a digital meter. This instrument typically consists of a signal-conditioning section, an analog-to-digital converter section, and a display section.

The signal-conditioning section converts whatever is being measured (ac, dc, voltage, current, resistance) into a proportional dc voltage. The analog-to-digital converter section, often referred to as the a/d converter, changes the proportional dc voltage to a digital signal that drives the display.

Regardless what type of meter is used, the following applications apply.

CURRENT MEASUREMENTS WITH AMMETERS

Ammeters are low-resistance instruments, so they can give accurate readings. Imagine inserting a flow meter into a pipe to measure flow; if the physical presence of the meter choked off most of the flow, the resulting reading would be worthless. The same thing is true with an ammeter; if its

Figure 12-2. Digital meter (Courtesy John Fluke Mfg. Co., Inc. Reproduced with Permission)

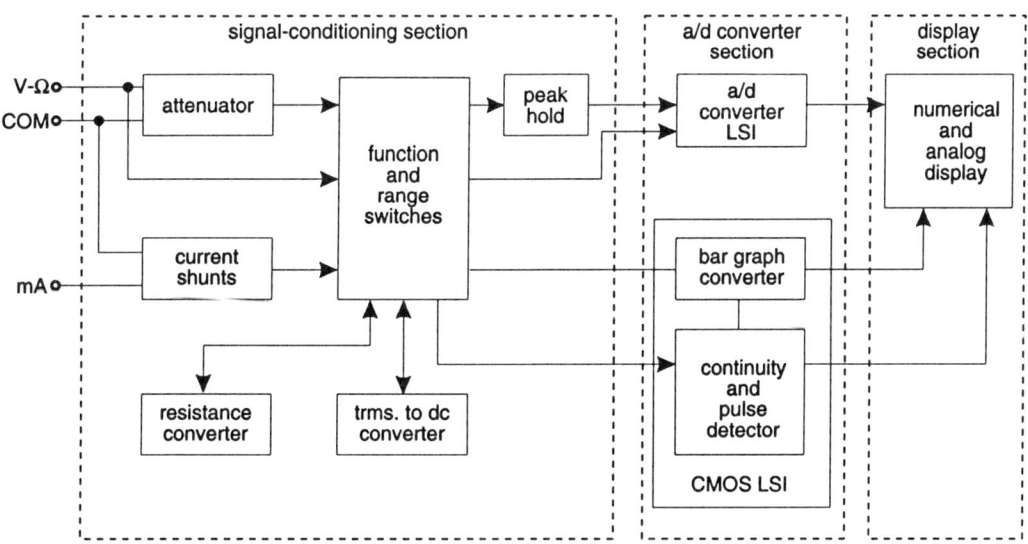

Figure 12-3. Block diagram of digital meter

Power Technology

reading is to be of any value, its internal resistance must be low, so it doesn't reduce the normal current flow.

Ammeters are always connected in series with the load. Connecting an ammeter across the line makes an instant short circuit. If power is applied, meter components are vaporized and personal injury can result.

The small-size wire used in the analog movable coil and the microelectronic circuits in digital meters limit internal current to the microamp and milliamp ranges. To measure larger currents, a **shunt** must be used. A shunt is a low-resistance bypass connected in parallel with the meter. It is designed so that a small amount of current goes through the meter, while the rest goes through the shunt. Figure 12-4 shows typical ammeter shunts.

If the meter movement has a **full-scale deflection** of 1 mA, and you wanted to double its capacity, use a shunt that would allow half the current to go through the meter, and half to go through the shunt. (Full-scale deflection indicates how much current is required to move or deflect the meter's pointer from zero to its maximum value; in other words, to move the pointer from the extreme left side of the meter's scale to its extreme right, Figure 12-5.) Now, instead of a full-scale deflection

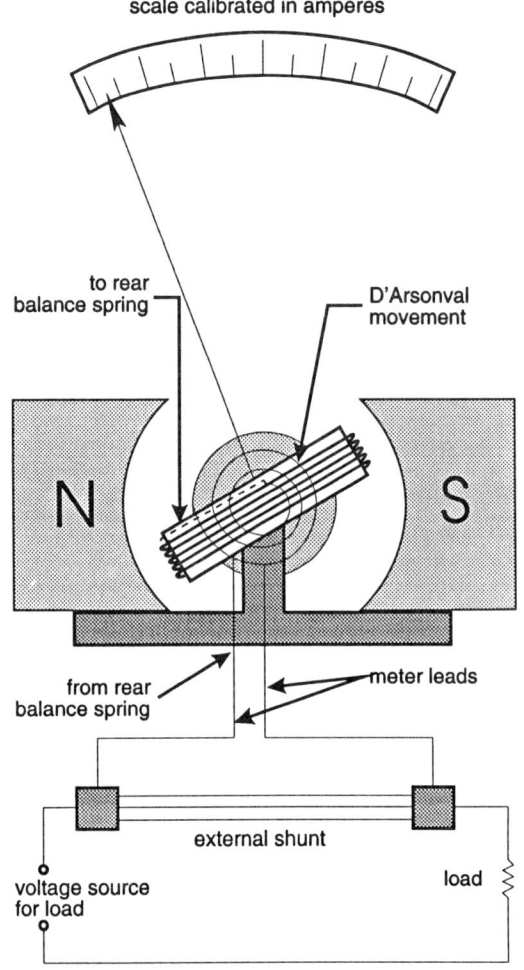

Figure 12-5. Full-scale ammeter deflection with externally mounted shunt

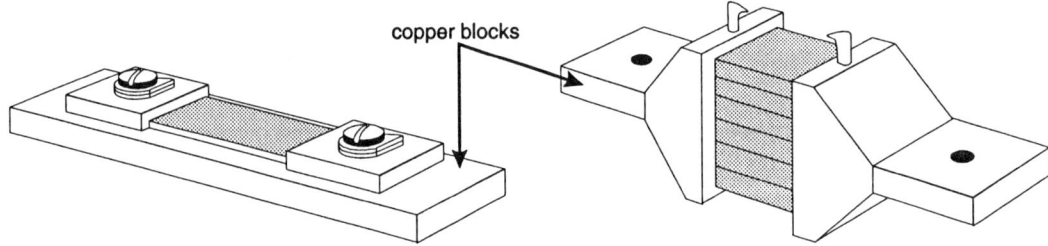

Figure 12-4. Typical ammeter shunts

equaling 1 mA, it would represent 2 mA. The lower the shunt resistance, the higher meter capacity becomes.

If the current to be measured is under 10 to 50 A, the shunt can be mounted in the meter case. Beyond that, the shunt starts to get large, requiring an external mounting.

CLAMP-ON AMMETERS

Now let's look at an ammeter that doesn't require any physical connections.

As current flows through a conductor, a magnetic field is produced. Amperage can be determined by measuring the intensity of this field. Clamp-on ammeters, Figure 12-6, and current transformers use this principle. The former is described here, the latter at the end of this chapter.

At times it is inconvenient or difficult to disconnect a load to take a current reading. Under these circumstances a clamp-on ammeter is used. All the ammeter requires to take a reading is for the clamp to encircle the conductor. Note that if it encircles both the hot and neutral conductor, their magnetic fields cancel out and the reading will be incorrect.

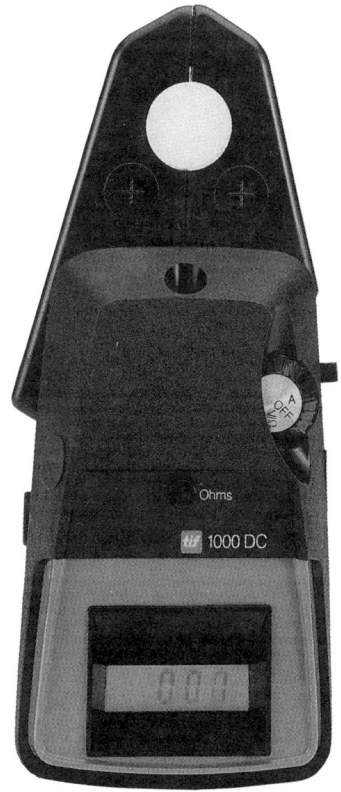

Figure 12-6. Clamp-on ammeter (Courtesy TIF Instruments, Inc., Miami, FL)

VOLTAGE MEASUREMENTS WITH VOLTMETERS

Unlike an ammeter, a voltmeter's resistance must be as high as possible. When a voltmeter is connected to measure the voltage difference between two points, it becomes part of the circuit. If the voltmeter has low resistance it draws a fair amount of current. This is no big deal when measuring power circuits, but bad news when measuring sensitive electronic circuits.

A low-resistance voltmeter lowers the effective resistance circuit and increases current flow, causing increased voltage drop and therefore resulting in an inaccurate low-voltage reading.

The sensitivity of analog voltmeters is expressed in ohms per volt and determined by dividing the resistance of the meter by the full-scale voltage reading. Inexpensive D'Arsonval meters are 1,000 ohms per volt, the better ones are 20,000 ohms per volt. Even the best meter movements require some power from the circuit under measurement to deflect the meter's pointer.

Power Technology

Electronic amplification circuits make meters more sensitive, because now the meter movement is powered from the amplifier, not the circuit under measurement. The amplifier's input can be designed into the megohm (million-ohm) per volt range. Like the first stereo sets, electronic meters used vacuum tubes and were called VTVMs (vacuum tube voltmeters). Modern instruments use solid-state technology, and instead of driving analog movements, they usually drive digital displays.

Electronic meters that use amplification circuits can be compared to the helpfulness of an amplifier to a person speaking to a large crowd. If the person needs to rely solely on shouting in order to be heard, the voice becomes strained and distorted, and many subtleties of normal speech are lost. If the person has a PA system, the amplifier does the hard work while the voice retains its normal inflections.

The same meter movement used to measure current is used to measure voltage. Instead of placing shunts in parallel, resistors are connected in series with the meter, Figure 12-7. The higher the voltage, the more in-line resistance is required.

AC MEASUREMENTS

If ac voltage or current were directly connected to a meter, the reading would be erratic; the analog pointer would just oscillate around zero, and the digital display would be a blur of unreadable numbers. In order to get accurate, meaningful readings, both types of meters must first convert ac to dc.

RESISTANCE MEASUREMENTS WITH OHMMETERS

An ohmmeter is basically a milliammeter with a dc source (usually a battery) and

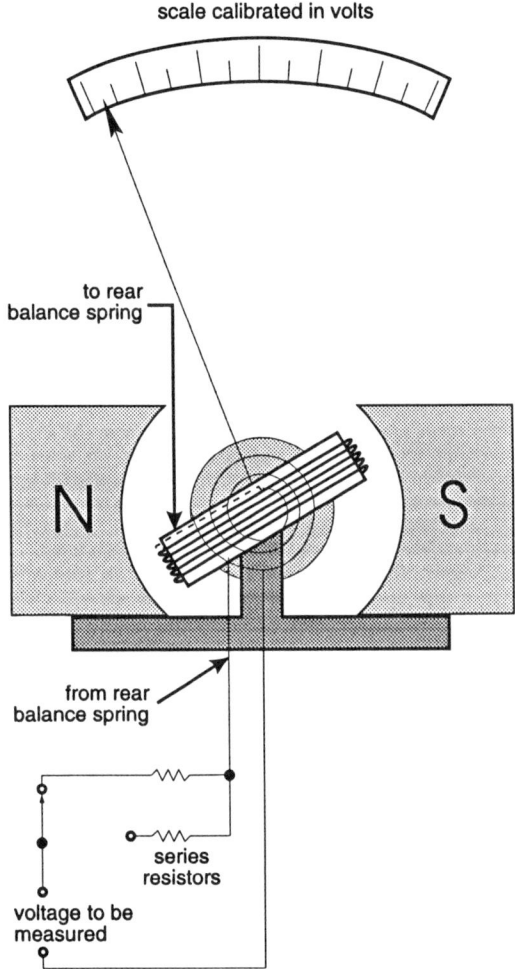

Figure 12-7. Voltmeter with resistors connected in series

resistors (one of which is variable, so the pointer can be zeroed), Figure 12-8. They have a measurement range from a few ohms to a few megohms.

When using an ohmmeter:

1. *Make sure there is no power on the device to be measured.* An ohmmeter has low resistance, and even low voltages can inflict terminal damage to the meter movement.

Electric Meters

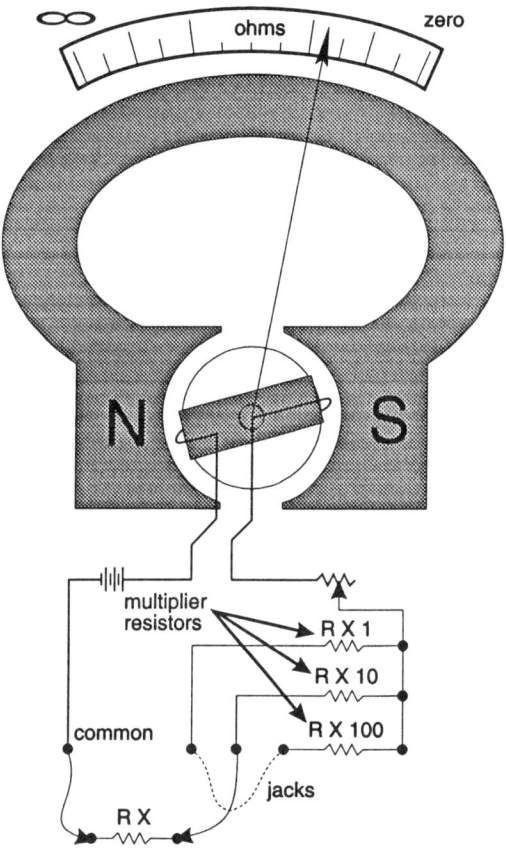

Figure 12-8. Typical ohmmeter

MULTIMETERS

Voltmeters, ammeters, and ohmmeters are sometimes integrated into a single package for greater servicing efficiency. Multimeters are capable of measuring up to 1,000 V, 10 A, and 2 megohms, Figure 12-9.

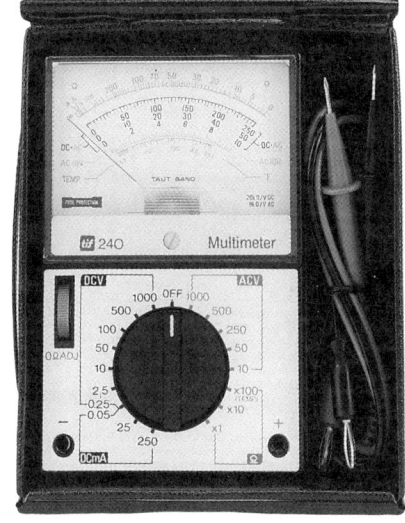

Figure 12-9. Multimeter (Courtesy TIF Instruments, Inc., Miami, FL)

2. Short the meter leads together; use the variable resistor to zero the pointer.

3. Connect the test leads in series with the circuit to be measured; read the scale. If the scale needs to be changed, the meter must be zeroed again.

When measuring in the higher-resistance ranges, don't let your hands become part of the circuit. At this level the meter is also measuring your body resistance in parallel with what is being measured. The result would be inaccurate low readings.

MEGGERS

To adequately test for insulation breakdown, it is necessary to supply a higher voltage than is furnished by an ohmmeter's battery. A megger is basically an ohmmeter with a hand-driven 500-V generator as its power source, Figure 12-10. Some new models are crankless; they are able to generate the required voltage electronically from batteries.

Power Technology

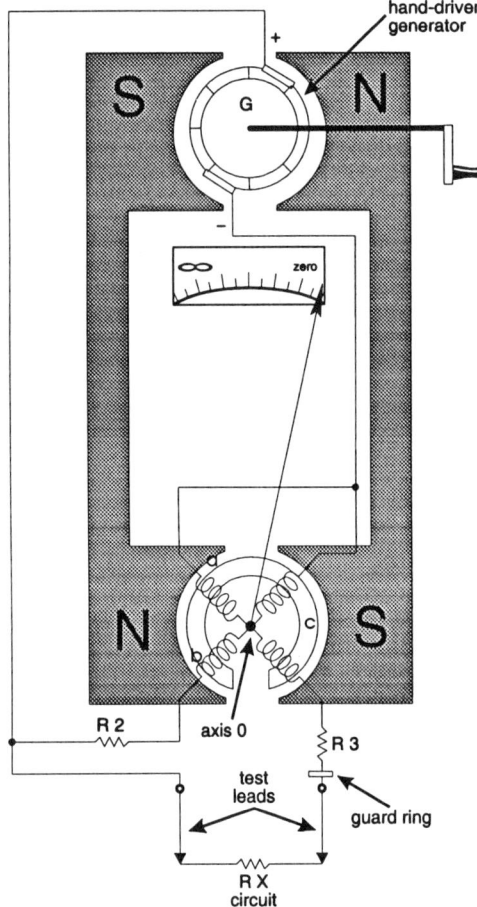

Figure 12-10. Megger with hand-driven generator

POWER MEASUREMENT WITH WATTMETERS

Both voltage and current must be measured to obtain power (watt) readings. To do this, a wattmeter uses an electromagnet instead of a permanent magnet, Figure 12-11. The strength of the magnetic field is proportional to current flow. The movable coil is connected as a voltmeter. The net result is pointer deflection proportional to the power consumed by the load.

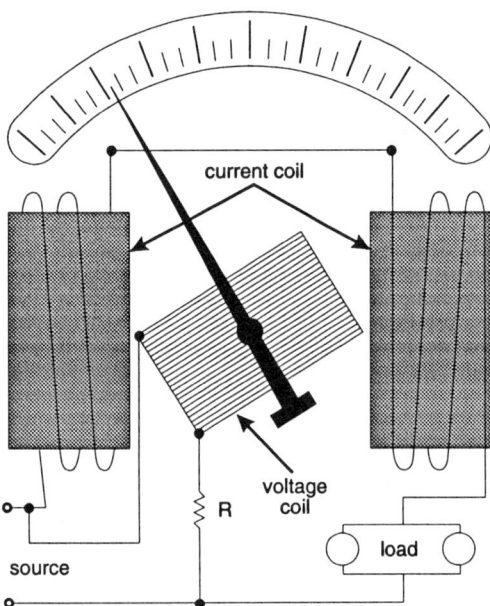

Figure 12-11. Wattmeter using an electromagnet

If the power is to be totalized, a watthour meter is used. In principle, the electromechanical watthour meter is a small motor whose speed is proportional to power passing through it. The shaft of this small motor drives a gear train which, in turn, positions indicating dials, Figure 12-12.

These indicating dials can be difficult to read when a pointer is directly on a number. In Figure 12-13, should the third dial be a 1 or a 2? Reading the dials right to left will make this interpretation easier. Since the fourth dial reads 9, it would seem logical that the third dial is a 1. If the fourth dial was beyond the 0, then the third dial would be at 2.

The meter in Figure 12-13 reads 0619.

Electric Meters

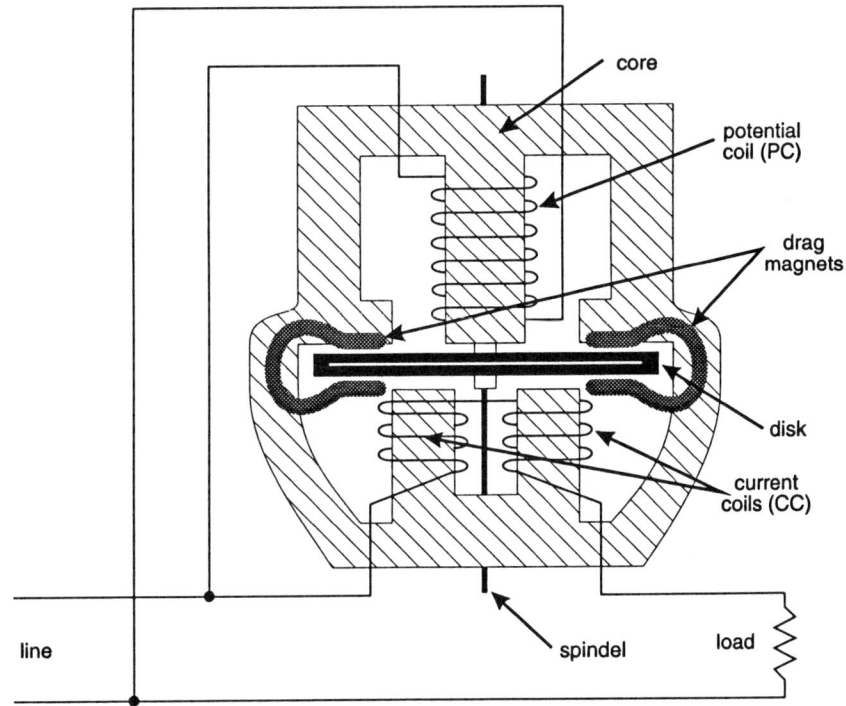

Figure 12-12. Watthour meter

Figure 12-13. Dials of a watthour meter

POTENTIAL TRANSFORMERS AND CURRENT TRANSFORMERS

Electrical instruments can only take so much voltage and current. In addition to physical limitations, there are safety considerations. It's just plain dangerous to have high energy close to operating personnel. Therefore, instrument potential transformers (PTs) and current transformers (CTs) are used to reduce voltage and current down to working levels before they are measured.

A PT reduces the voltage to be measured down to 150 V. They come in different ratios, so if the primary is 14,400 V or 144,000 V, the secondary is still under 150 V. Figure 12-14 shows a PT reducing 14,400 V down to 120 V. The scale on the 150-V-reading voltmeter is calibrated so that it reads 14,400 V.

Current transformers reduce current down to 5 A, Figure 12-15. They too are available in different ratios, so any current level can be safely measured. Because a current transformer steps down current, this means voltage is stepped up. This isn't a problem as long as a meter is connected to the secondary.

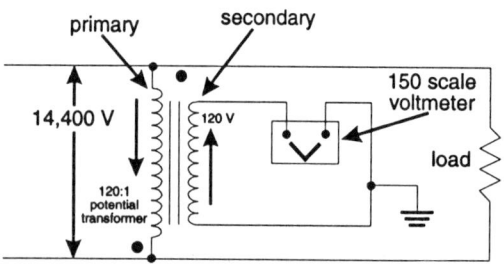

Figure 12-14. PT reducing 14,400 V down to 120 V

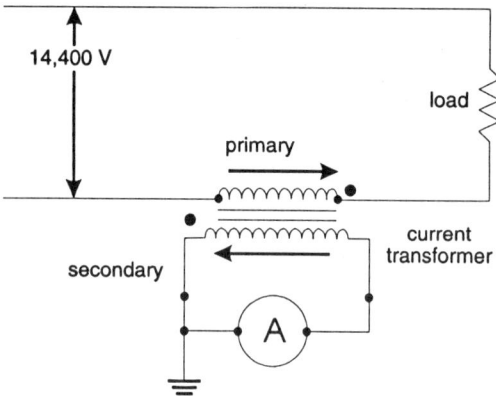

Figure 12-15. CT reducing current down to 5 A

Because a current transformer is a low-power device, the meter loads it down safely and efficiently. However, there is a big problem if an energized current transformer secondary is opened. There will be dangerously high voltage across the secondary. Excessive core loss will overheat the transformer, and the transformer and cable can be damaged.

The secondary of a current transformer should not be open-circuited under any circumstances.

Pictured in Figure 12-16 are the connections of a PT and CT to a single-phase wattmeter. The dots in the drawings are polarity markings, so the meters can be connected for proper phase relations.

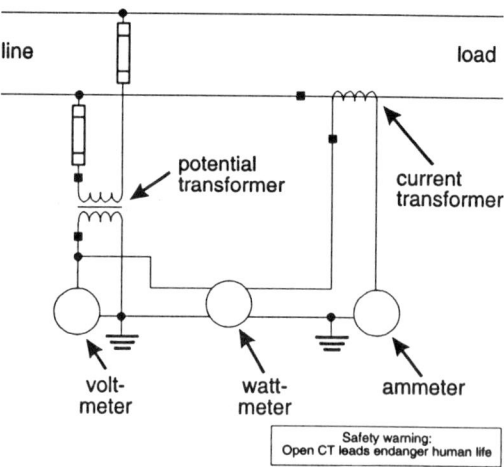

Figure 12-16. PT and CT connections to single-phase wattmeter

NOTES

[1] This quotation was excerpted from John Bartlett's *Familiar Quotations*, Fifteenth and 125th Anniversary Edition, published by Little, Brown and Company.

Chapter 13
ENERGY MANAGEMENT

AUDITS, SUB-METERING, CONSERVATION

The ability to "manage" a non-tangible substance such as energy may seem a bit farfetched. It can, however, be done; and our ability to do so depends on our ability to measure energy. Consider these gems of scientific thought:

"If you can't measure it, you can't manage it."

"If you can't count it, you can't control it."

These two statements may not apply to every facet of our lives. However, when discussing energy management, they ring true.

ENERGY AUDITS

For a commercial building, the best place to start measuring energy is with an energy audit. As with any audit, the more information you can gather, the better and more accurate the results.

Below is a brief outline describing how to conduct an energy audit:

1. Define scope — Identify needs and objectives.

2. Gather data — Assemble available records. Measure to fill gaps in data.

3. Analyze data — Identify areas of high energy cost.

4. Initiate projects — Propose and justify energy projects. Design and implement projects.

5. Monitor energy use — Verify performance and savings. Continue monitoring key locations. Adjust and/or modify for desired results.

This discussion will be limited to electric power audits.

What type of building are you most involved with: An office building, factory, department store? Whatever it is, pick your initial target — the building that you think will yield the highest energy-conservation return.

Start gathering data. One year's supply of monthly electric bills is a good start. They contain a great amount of information regarding on-peak energy use, off-peak energy use, demand, and power factor.

Energy bills need additional collaboration to get a true picture of costs. If analyzing building comfort systems is your objective, you will need weather data, especially degree days, to give you a better picture of where those utility charges came from.

The **walk-through audit,** the least costly, identifies preliminary energy savings. A visual inspection of the facility is made to determine maintenance and operational energy-saving opportunities, plus the collection of information to determine the need for a more-detailed analysis. Most power companies will help conduct walk-through audits.

Mini-audits require tests and measurements to quantify energy uses and losses, to determine the economics for change. If the load profile isn't available from the power company, you need to arrange your own mini-audit. Sophisticated equipment isn't always required; as stated before, a recording ammeter connected to the incoming service gives some indication of energy-use peaks and valleys.

Maxi-audits go one step further. They evaluate how much energy is used for each function, such as lighting, process, hvac, etc. These audits also can aid computer modeling, to determine energy-use patterns and predictions on a year-round basis. Variables such as weather data can be programmed into the evaluation.

LOAD PROFILES

The next step is to get a load profile of your facility. No meaningful study can be done without one. Load profiles show the peaks and valleys of when power was used during the day. The profile might point out that, at 1:00 p.m. every day, there is a large spike that constitutes a large percentage of your demand charge.

Could that spike represent the maintenance department checking out a machine? Could this maintenance be rescheduled? Also, are machines running idle during the peak period? It would be better to turn them off. Remember, it isn't the initial start that increases demand; it's the running time.

Some monthly electric bills are recorded on tape at the meter and picked up every month by the utility for processing. If your demand is recorded on tape, then the power company can supply, for free or a small charge, the building's monthly profile. Some utilities even break it down by week and have a separate chart of the peak day.

If profiles are not available from the power company, produce them yourself. Recording wattmeters are a bit expensive, but these instruments can be rented. At least, a simple recording ammeter clamped to only one phase of the incoming service will give you *some* idea of power usage quantities.

What if the curve on your peak demand day looked like the curve on Figure 13-1? Where did that spike come from? The thin spike is about 25% of your demand; it cost you a bundle of money. If the cause can be identified and managed, future bills won't look so bad.

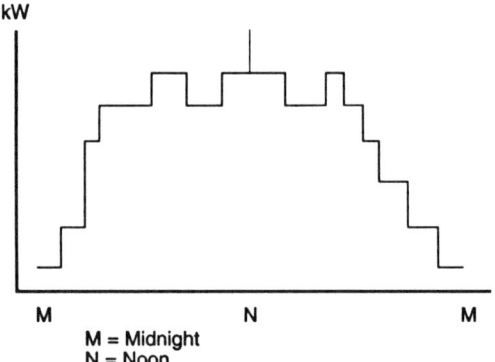

Figure 13-1. Peak demand usage; 25% of demand costs from one little spike

What if the curve on your peak demand day looked like Figure 13-2? There's not much choice; you need a lot of hard work to bring the overall level down.

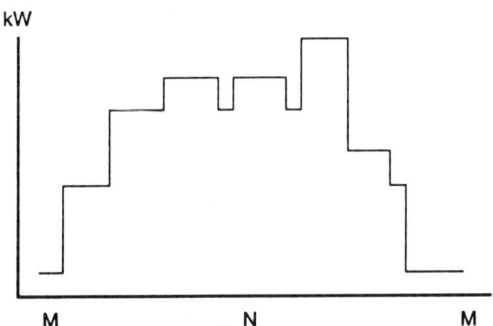

Figure 13-2. Fairly level peak demand usage

SUB-METERING

Many people think that sub-metering energy use doesn't really make a difference. For those unbelievers, this is an account of two new, near-identical apartment complexes with identical electrical service capacity. One had no problems; its usage was as predicted. The other had big problems: The lines were overloaded, and there was even a transformer failure.

After investigating, it was found that the complex with no problems had individual metering. Tenants were responsible for paying their own electric bills. As a result, air conditioning wasn't operated during the day, when apartments were empty, and thermostats were set at reasonable levels.

In the other complex, utilities were included in the rent. It was found that air conditioners were on full blast all the time because the tenants saw no penalty for squandering energy. Having energy costs rolled into the complete rent package was almost like saying it was "free." There was no accountability. Usage was very high. The landlord was taking a bath.

Human nature being what it is, if people are not charged for something, more of it will be used. If each operation was metered and charged accordingly, energy costs would go down. Efficiency also might increase, because energy consumption per unit of production or service can be tracked and analyzed. Inefficient equipment can be repaired or replaced, and inefficient operations can be improved.

Technology has reduced the cost of sub-metering. Instead of needing to install large, heavy-duty, glass-encased meters like utilities must do, individuals can easily install small, electronic-based meters. The sometimes hard-to-read circular dials have been replaced with digital readouts.

Data collection can be as simple or as complex as the situation demands. It can vary from someone taking a monthly walk and reading the meters, to the data being gathered minute by minute and fed into a personal computer spreadsheet for analyses.

By keeping a minute-to-minute eye on energy usage, demand charges can be reduced. If you find that the top 10% of demand charges happens during one 2% or 3% period of a 24-hour day, find out what caused those spikes. Once identified, action can be taken to eliminate them.

Several manufacturers offer equipment and software packages for sub-metering and load profiling. The data-gathering computer doesn't even need to be on site. With modems, the monitoring and evaluation can be done at a remote location.

> **Why meter?**
>
> - To be able to charge to individual departments
> - To increase accountability for energy used
> - To be more responsive to the efficiency of equipment and systems (i.e., chillers)
> - To provide information for audits of energy projects
> - To facilitate maintenance work, identify performance problems, and increase feedback to managers
> - To identify potential future additional energy savings

CONSERVING ENERGY

The first rule of energy conservation is, *"If you're not using it, turn it off."*

The most obvious equipment to turn off are the lights. One argument against this is that the life of a lamp is shortened every time it is turned on. True, but the energy that lamp consumes is much greater than its initial cost. A good rule of thumb is, if the lamp isn't going to be used within 15 min, turn it off.

Another argument is that the initial surge of energy required to start lights and motors wipes out any potential savings. This initial surge is minuscule compared to the amount of energy saved.

In areas such as conference rooms and lavatories, motion detectors can be effective. As soon as the demand for these facilities ends, out go the lights. Installing motion detectors in offices is becoming more common, but it must be evaluated on an individual basis.

Photocells have been around for awhile. These do an excellent job of turning lights on or off in direct response to the amount of natural light available.

With personal computers so common, their energy usage rivals the light bill. It would be nice if at least the screen would turn off automatically during periods of inactivity.

The second rule of energy conservation is, *"Don't do it twice."*

For example, if a corridor is fully lit with sunlight, turn off the lights. If it is 20°F outside, see what can be done to cut back on the operation of the chiller plant.

Some energy-conservation measures require considerable up-front investment. For example, some capital is required to correct hvac systems that were originally designed to reheat mechanically cooled air, for more-sensitive temperature and humidity control.

ENERGY MANAGEMENT SYSTEMS: PAYBACK, AUTOMATION, AND DESIGN

Energy management systems can be anything from a simple time clock, to a computer-based control system that does everything from operating the parking lot lights to deciding to burn oil or gas in the boiler.

It's generally best to keep systems as simple as possible, but not so simple as to deny your facility of energy savings that could be achieved through a more-sophisticated system.

PAYBACK

Going through the exercise of determining the economic justification of a project is always worthwhile. Companies have different criteria to determine if it makes sense to proceed with a proposal, but most use the simple payback calculation for the first go-round.

$$\text{simple payback in years} = \frac{\text{cost of project}}{\text{yearly savings}}$$

If the calculated payback comes in at under two or three years, this is evidence that it is worth the effort to proceed with a more-involved payback calculation.

AUTOMATION

A lot of operations are already automated. Chillers regulate water temperature, boilers regulate steam pressure, and thermostats regulate room temperatures. It would be unthinkable to assign people to these tasks.

The environmental conditions of buildings are continuously changing. Humans simply cannot monitor, evaluate, and act upon all the inputs required to operate a plant at top efficiency as well as computers can.

Here are some of the questions operators need to entertain when operating a facility on the razor's edge of efficiency:

- When should a chiller be taken on- or off-line to match load conditions?
- Exactly how much cooling tower capacity is required to minimize chilled-water plant costs?
- What is the temperature drop across the cooling tower?
- How many degrees above wetbulb is the tower operating?
- Are the proper amount of cooling tower cells on-line?
- Can the condenser water flow through a chiller be reduced and still maintain chiller efficiency?
- What is the efficiency of the chillers in tons per kW?
- Has chiller performance been falling off lately due to fouled tubes?
- Do all the chillers have the proper chilled-water flow rate?

Can any one person process all the required inputs and arrive at an optimized solution 24 hours around the clock? Only with microprocessor-based controls is it possible to collect, process, evaluate, and act upon hundreds of pieces of information.

The purpose of automation is not to get rid of the operators. Its function is to free the operators for other work, such as preventive maintenance and troubleshooting.

If you get into automation, you will be surprised just how little information the operators have to work with when running the plant. Since computers require information on all aspects of the plant, automating forces the installation of the proper instrumentation (to measure temperatures, flows, power consumption, etc.).

And the correct software is where it's at. The hardware is useless without it.

DESIGN OPTIONS

Rule three of energy management is, *"If you can, do it at night."*

Demand charges can account for one-third to one-half of the cost of power used during the day. By shifting high energy consumption to off-peak hours, you can avoid those high demand charges.

Consider shifting production schedules and applying thermal storage technology. (Thermal storage will be covered in a later chapter.)

DEMAND LIMITING

If loads can't be shifted to off-peak rates, try to schedule activation times so that demand is level. Interlock large loads so they don't operate simultaneously.

Don't test large loads during peak usage. You could be testing a pump or compressor just to see if it works, and then get hit with a large demand charge at the end of the month.

Cooperation between departments is essential. See if there is an application for computer control.

DUTY CYCLE

Don't try to short cycle to save energy. Some hvac manufacturers void their warranty if they find their unit was connected to an energy management system. Best to check before.

Motors, especially the large ones, will be damaged if started too many times during a short period of time. A duty cycle program needs to be smart enough to not damage equipment.

MOTORS

Most motors are oversized and continuously operating at less than full load. Motor efficiency under these circumstances is lower than expected, because the manufacturer's published efficiencies are based on full load.

To determine the percent of full load a motor uses requires special equipment. Figure 13-3 gives a good approximation by using a simple ammeter. By taking an ammeter reading and comparing it against the motor's rated full load amps (FLA), the efficiency and percent of full motor load can be determined. The motor in this example is loaded to 53% of its full load amps. Figure 13-3 shows that it is only loaded to 40% of its full load and is operating at 73% efficiency.

Would it pay to replace this motor with one half its size? At 80% of its full load, its efficiency increases to 88%. Would a 15% increase in efficiency justify a motor changeout? One of the main items to consider is how many hours a year it operates. If it just runs a few hours a day, then the payback period would take a long time. If it operated around the clock, the payback would be very attractive.

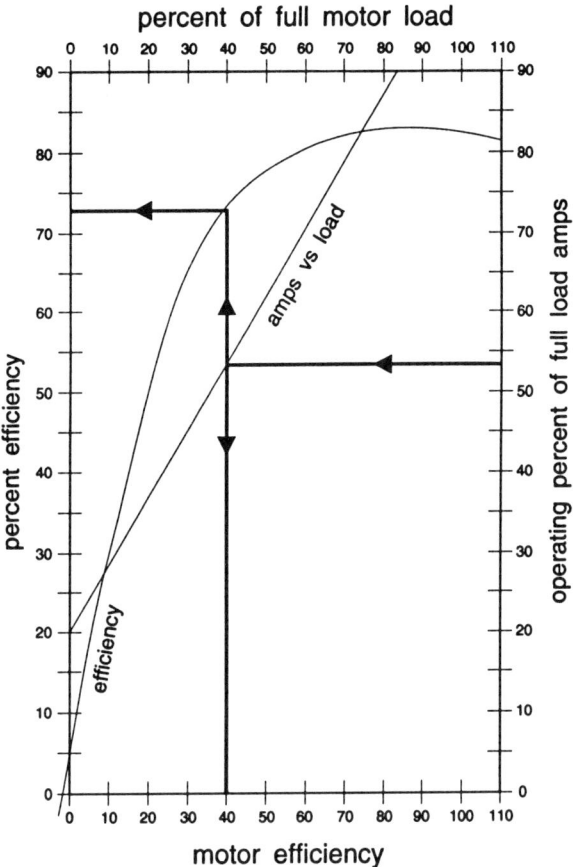

Figure 13-3. Percent of full motor load

Chapter 14

THERMAL STORAGE

Thermal storage is a technique that shifts a portion of the electrical load from peak daytime hours to off-peak nighttime hours, to reduce demand charges and thus lower the electric bill. Both heat and cold can be thermally stored; however, we will concentrate solely on cold storage, which is more commonly used in commercial buildings.

ECONOMIC BENEFITS

The economic attraction of thermal storage is made possible by electric utilities' demand charges and time-of-day rates. These rates are typically highest during normal business hours, when air conditioning and other electrical uses are at their peak. One way to reduce these charges is to run a thermal storage system at night, when electricity is less expensive, to make ice at night for air conditioning use during the following day.

Thermal storage is not a new concept. A system was installed in a Dallas, Texas hotel as early as 1925. The emphasis in these early applications was to minimize the initial cost for a short-term cooling load.

Churches, theaters, and process applications such as dairies have been good candidates for thermal storage. For example, a church might have a 50-ton, five-hour, one-day-a-week cooling load, with a total of 250 ton-hours of cooling capacity required. Rather than purchasing a 50-ton system that would only operate five hours a week, it makes more sense to install a 5-ton system to operate and store cooling for 50 hours. The same total cooling capacity (250 ton-hours) is produced, and system cost is substantially reduced, even when the cost of the storage equipment is included.

Some refrigeration terms:

- A **ton** of refrigeration is equal to the removal of 12,000 Btu per hour.

- A **Btu**, or British thermal unit, is the amount of heat required to raise 1 lb of water 1°F. In the case of refrigeration, it's the amount of heat removed from 1 lb of water to lower it 1°F.

- A **ton-hour** is the product of tons produced over a period of time, and is analogous to the kilowatt-hour in electricity.

More recently, air conditioning systems in office buildings, schools, institutions, laboratories, large retail stores, libraries, and museums now are benefiting from the economic benefits of thermal storage systems.

There is some potential to save money through energy reductions, as well as on utility charges. If the small air conditioning plant bolstered by thermal storage is operated at its most-efficient point, it still probably uses less energy than a larger system, which often operates far below its rated capacities. Therefore, money is saved two ways: when demand charges are reduced, and when less capital is needed to operate the smaller plant.

Shown here are two building load electric demand profiles, Figure 14-1. The first shows how the cooling load just about doubles at peak demand. The second profile shows the same building, but with the cooling demand shifted to off-peak hours. The demand portion of the electric bill has been considerably reduced.

Unfortunately, the thickness of your wallet also would be considerably reduced to pay for this large thermal storage installation.

As is commonly the case, it costs a few bucks to save a few bucks. Therefore, several options must be evaluated.

The **full-storage** option, Figure 14-2, is the most expensive to design and install. Figure 14-2 also shows the relative size of refrigeration plants and storage required for the most common thermal storage systems.

Partial-storage systems run more hours than full-storage systems, so less demand-reduction is obtained. However, they are initially less expensive to install than full-storage systems, because less storage capacity is required and smaller-capacity refrigeration equipment is used.

The process of shifting a load from a high to a low demand period is called load leveling. This scheme minimizes the compressor's required size and significantly reduces the building's peak demand. When this strategy is used, equipment is sized so the refrigeration equipment runs all day.

The overall effect of this scheme is to level the cooling component of the building's load, Figure 14-3. During peak hours, part of the cooling load is met directly by the

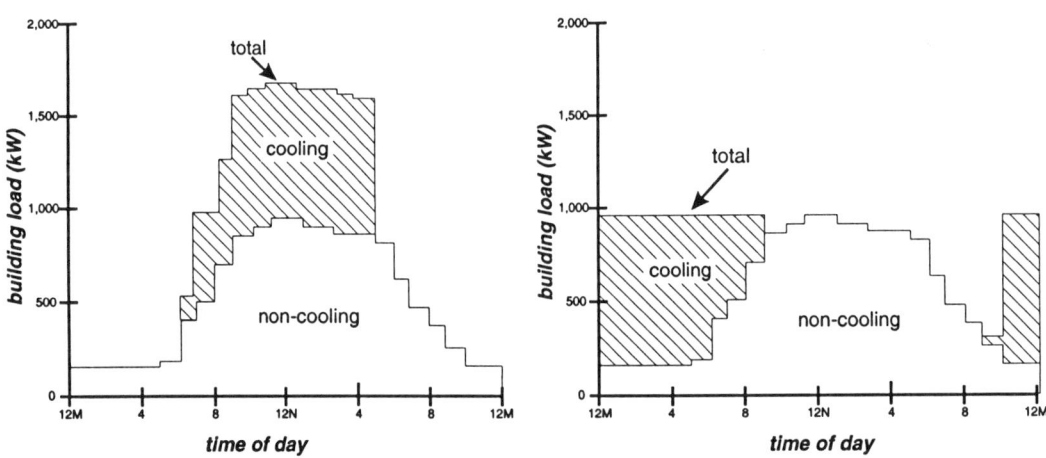

Figure 14-1. Two building load profiles, with and without load shifting

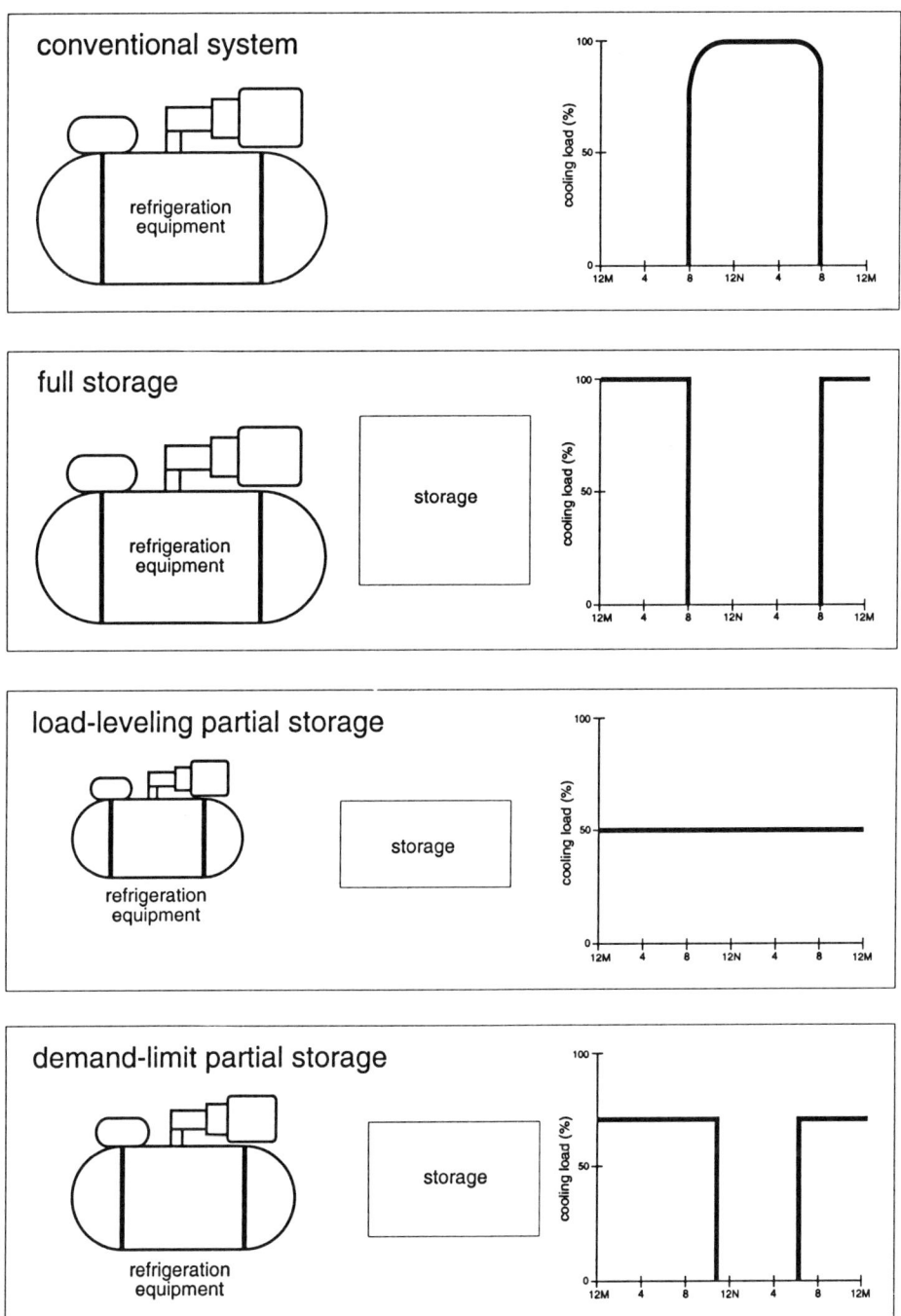

Figure 14-2. Relative size and storage for common thermal storage systems

compressor and partially by storage. The storage must be adequate to supply all of the building cooling load not met directly by the refrigeration equipment. Here, about 60% of the building's peak-hour cooling load is supplied from storage.

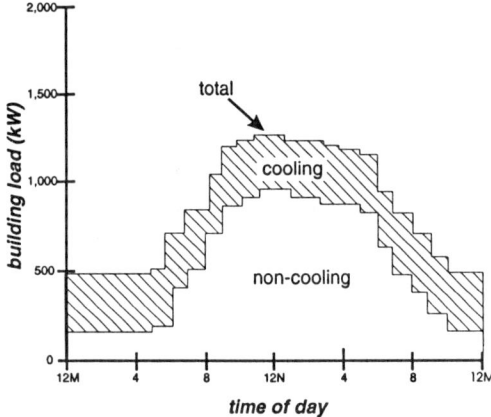

Figure 14-3. Example of load leveling

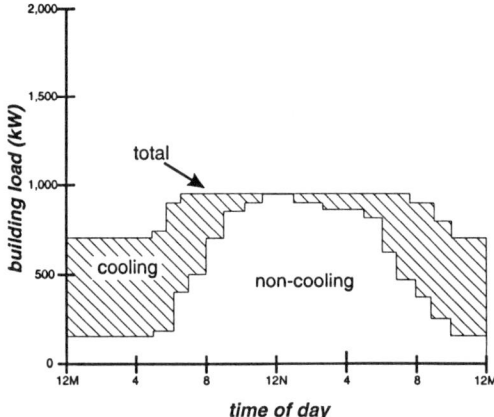

Figure 14-4. Peak electrical demand reduced to non-cooling peak demand

Demand-limited systems use sophisticated techniques to control the compressor, thus preventing its operation during peak periods. At all other times, the compressor is operated to meet both the direct cooling loads and to cool the storage medium.

As you can see, the building's peak electrical demand has been reduced to the non-cooling peak demand, Figure 14-4. Instead of displacing the cooling load entirely to the off-peak period as full storage does, a considerable amount of the cooling load is fit into periods when the non-cooling load is low.

In most situations, thermal storage does not save energy; it just shifts the time of consumption. However, it can be argued that thermal storage leads to more-efficient operation. Conventional systems operate at part-load most of the time — not the most efficient way to operate air conditioning equipment. Cool storage equipment operates at full capacity, significantly increasing efficiency.

In addition, efficiency is improved at night when outdoor air temperatures are cooler, resulting in improved condenser heat rejection. Also, cooler night air can be more economically processed with thermal storage systems. (With cooler air, less air needs to be delivered for the same amount of cooling. And less air means smaller ducts and fans.)

UTILITY PERKS, INSTALLATION COSTS

Installing a thermal storage system can be a win-win situation for your building and local electric company. Many utilities offer financial incentives for buildings to install thermal storage systems, to help postpone their own need for expensive generating stations which, to meet high peak demand, sometimes need to be built.

There is also an environmental cost for utilities to meet peak demands. During peak times, the older, more-inefficient generating station units are brought on-line to maintain service.

More good news: Thermal storage permits the use of smaller, less costly refrigeration compressors, whose reduced electrical load can result in a lower-cost power distribution system.

In addition, when ice is used as the storage medium, a lower supply water temperature can be economically produced. The lower water temperature means a greater temperature difference exists between it and the air to be cooled. This provides an opportunity for lower-temperature air distribution, which means less air needs to be circulated to achieve the same cooling. And, as mentioned above, with less cfm (cu ft per min) of air required, smaller air ducts and fan motors can be used.

Sobering news: Costs run relatively high to provide the storage medium. It can be in the form of water, ice, or the more exotic eutectic salts that freeze at temperatures greater than 32°F.

TYPES OF THERMAL STORAGE SYSTEMS

There are three basic types of thermal storage system designs: ice storage, chilled-water storage, and eutectic salt systems.

ICE STORAGE SYSTEMS

Here is a diagram of a typical ice storage system, Figure 14-5. The refrigeration system's job is to make ice in a tank. The chilled-water pump circulates water around the ice to cool it down, then pumps this chilled water out into the building to cool down inside temperatures.

CHILLED-WATER STORAGE SYSTEM

Figure 14-6 shows one of several arrangements for a chilled-water storage system.

Chilled-water storage takes up more room than does ice storage. Water storage, however, has this advantage: Existing chillers can be integrated into the new scheme. There are several ways to keep the warm and cold water separated. This system uses a flexible diaphragm. Other systems use baffles. And large tanks need no separation schemes, since their design utilizes the fact that hot and cold water stratifies naturally.

Now, why is less space required for ice? It all has to do with how water freezes. To fully understand this process, we need to define sensible and latent heat.

Sensible heat can be detected by a thermometer. Fill a pot with 1 lb of warm water, stick a thermometer in it, then put it in the freezer. Every 5 min or so, note the temperature; it is dropping, as expected. Stick your finger in the water. You can sense it getting colder.

Something unexpected happens when the water reaches 32°F. The temperature stabilizes — it doesn't drop any more. But the freezer is still colder than the water, so the freezer must still be removing heat from the water. In fact, the water is slowly changing to a solid (ice). During this period, **latent heat** is being removed. (Latent means hidden, like latent finger prints.) Up until this point, to reduce the temperature of 1 lb of water 1°F, one Btu had to be removed. Once the water's temperature reaches 32°F, to convert this pound of water into a pound of ice, 144 Btu — the latent heat — must be removed.

This is the advantage of ice. Plop a few cubes into your tea. Depending on the tea's temperature, the ice cools it down and some

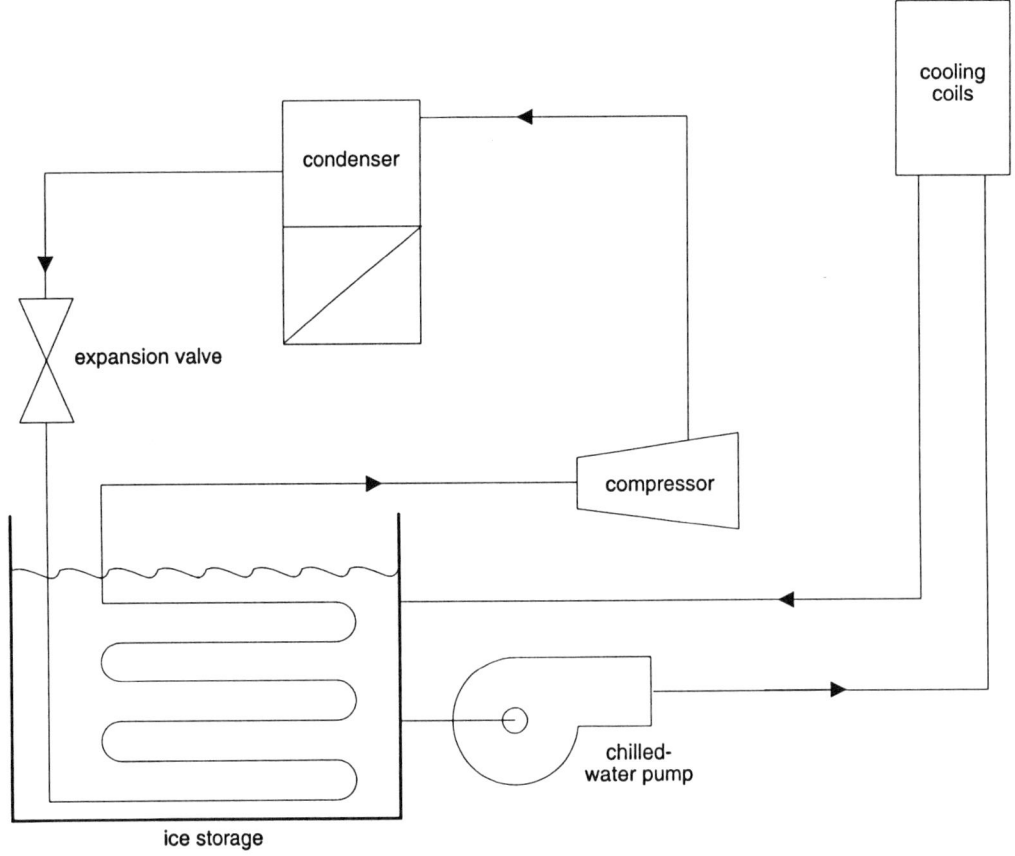
Figure 14-5. Diagram of a typical ice storage system

of the cubes remain. To melt 1 lb of ice cubes, the tea must absorb 144 Btu. Therefore, because the melting process absorbs a lot of heat, ice systems take up less space than water-storage systems.

EUTECTIC SALT SYSTEMS

When mixtures of inorganic salts and other additives are added to water, the freezing point of the mixture is raised — 47°F is common. The advantage of eutectic salt thermal storage systems is that existing chillers can freeze the storage medium, and the storage volume can be reduced because the principle latent heat can be utilized.

TOUGH CHOICES

Sometimes thermal storage projects can dovetail with other work. For example, do you have, or are you going to build a fire protection water reservoir? Often these water storage facilities, such as water towers, can also serve as the thermal storage reservoir.

With so many options, you need to use a detailed procedure to find the right thermal

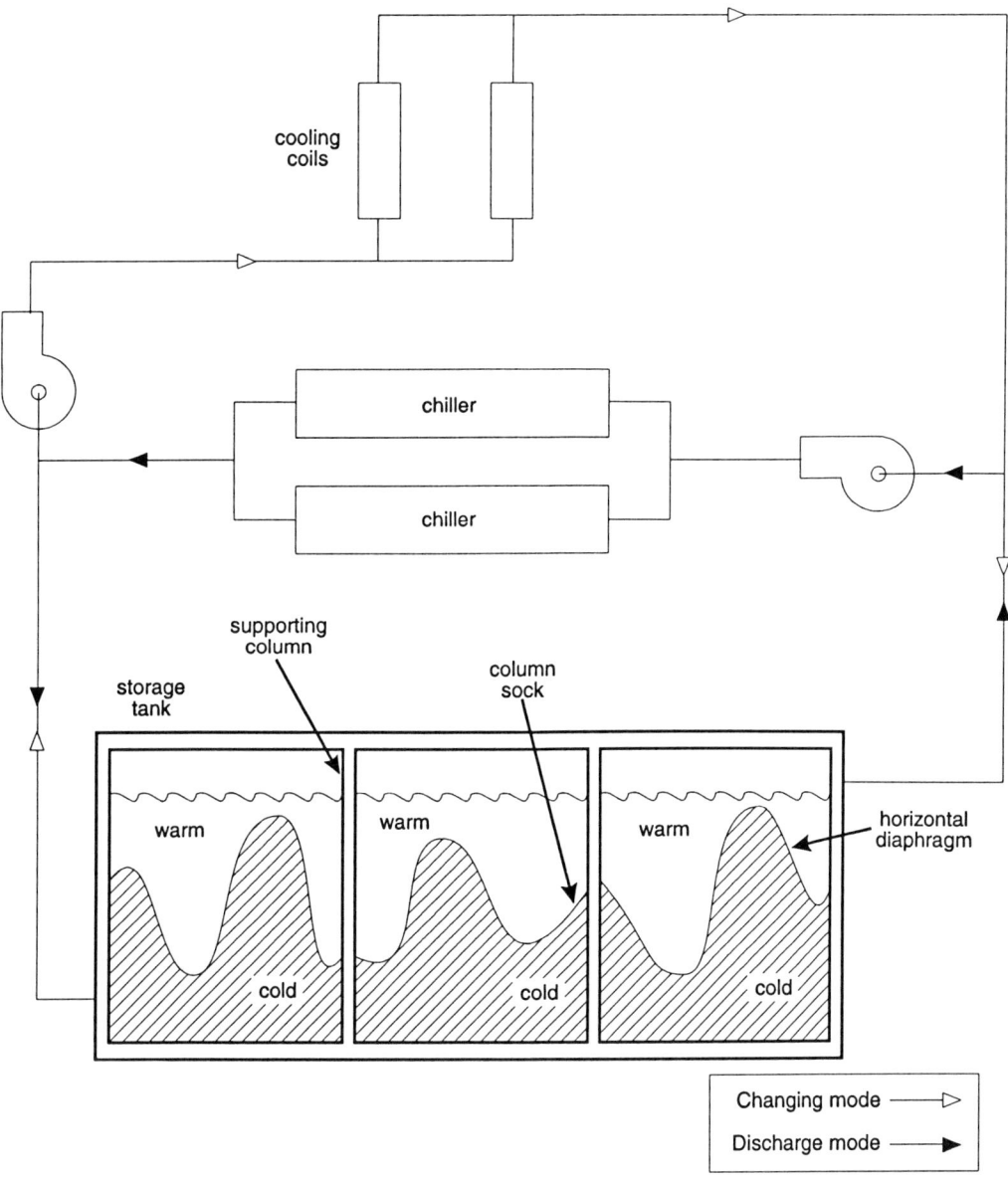

Figure 14-6. Chilled-water storage system

storage combination. All have their advantages and disadvantages. To help weed out all the various combinations, contact reputable engineering firms; have them come in and give their presentations. Also, visit existing installations and ask questions. If the thermal storage concept seems suitable for your situation, then professional analysis might be appropriate.

Chapter 15

SYNCHRONOUS GENERATORS, INDUCTION GENERATORS, AND THE REFRIGERATION CYCLE

Electric power is generated by the relative motion of a conductor passing through a magnetic field. It doesn't matter if the magnetic field or the conductor are moving, as long as relative motion exists between them. The conductor in which the electric power is generated is called the armature. Electricity is necessary to power most of today's refrigeration and air conditioning systems.

Generators are similar to motors. In fact, generators can act as motors, and vice versa. Generators come in two types: synchronous and induction.

SYNCHRONOUS GENERATORS

Synchronous generators, by far the most common, can generate electricity on a free-standing basis (unlike induction machines, which need to be connected to an energized line in order to work).

The ac generator in Figure 15-1 shows a rotating armature cutting through magnetic lines of force.

It's the field that rotates in actual synchronous ac generators, Figure 15-2. The rotating magnetic field in the rotor cuts across stationary armature windings in the stator.

Output power is generated in the armature. This important feature makes large, high-capacity ac generators possible. Because the armature is stationary, its output can be directly connected to its load.

With dc generators, the field is stationary and the armature rotates. The armature's output cannot be directly connected to its load; it first must go through brushes of a limited capacity. You've probably seen sparks flying off the brushes of a drill motor. Just imagination the design and maintenance problems associated with large-megawatt dc machines!

Whenever rotating and stationary parts make direct contact, one element is made from a hard material and the other from a soft material. Take, for example, engine and transmission shaft seals: When electricity must be transferred between stationary and rotating parts, carbon brushes are used. This is because carbon is a good conductor and has good wearing characteristics when rubbed against a hard smooth surface. However, they do wear and have limited current-carrying capacity.

Like synchronous motors, synchronous generators require excitation for their operation. Small, portable generators use permanent magnets built into the rotor for their excitation. Larger machines, which need

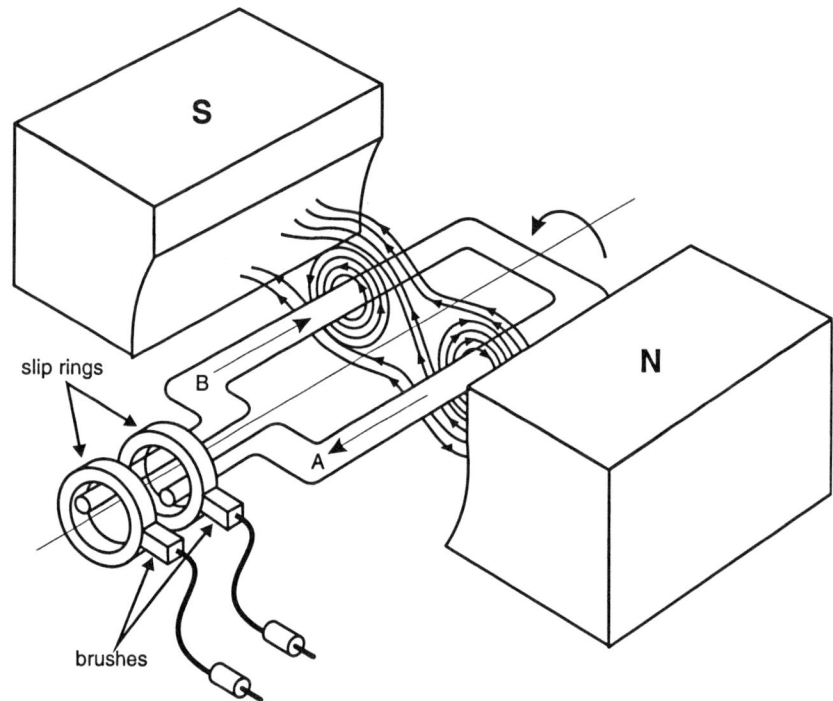

Figure 15-1. Ac generator's rotating armature cutting through magnetic lines of force

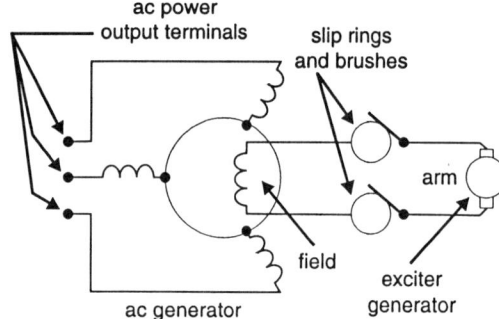

Figure 15-2. Rotating field in synchronous ac generator

stronger magnetic fields, use electromagnet windings built into the rotor for their power. Excitation power can come from dc sources such as batteries, rectifiers, or dc generators.

This power can be fed to the rotating magnet via slip rings attached to the shaft and stationary brushes (just like synchronous motors). With large machines, the exciter can be mounted directly on the generator's shaft.

All generators, whether synchronous or induction, must be cooled. The smaller ones are cooled with air, the larger ones with hydrogen (H_2). At a 100% concentration, hydrogen is quite safe and has several advantages over oxygen for these applications. First of all, hydrogen is less dense than air, so rotor **windage** (friction loss) is reduced. The second advantage is that electric arcs are quenched quickly in hydrogen. Finally, hydrogen-cooled windings tend to last longer than air-cooled ones. Forty years is not uncommon.

The frequency output of a synchronous generator is directly proportional to its rpm. The prime mover's speed must be well regulated in order to maintain stable operation.

As you can expect, all emergency generators that are expected to produce power during utility company power outages are synchronous machines.

Paralleling is the way in which a generator is placed in service on a system that is already energized. Paralleling synchronous generators when additional capacity is required can be tricky. The output voltages of both generators must be in phase before the connection is made or damage will occur. "In phase" means that the voltages are equal. When in phase, the generator being brought on-line will direct its output into the system; when done out-of-phase, the equivalent of a short circuit occurs. The resulting high short-circuit current could not only damage equipment, but also kick the entire system off-line.

Picture two ships attached together by a sturdy line, bobbing and rolling in a rough sea. If they bob and roll together, no great strain is put on the line. But if one zigs while the other zags (a.k.a., out-of-phase), the line is put under a great deal of strain. In this type of situation a line may part, or a fitting on one of the ships would break free.

Various schemes and instruments — from simple light bulbs to sophisticated instrumentation — have been devised to ensure that paralleling goes smoothly. Paralleling can be done manually or automatically.

One method of paralleling is shown in Figure 15-3. When approaching synchronism, lamp L1 will be dark because the "C" bus voltages will be close. When lamps L2 and L3 are the same brilliance, then generators are closer together and correct rotation is verified.

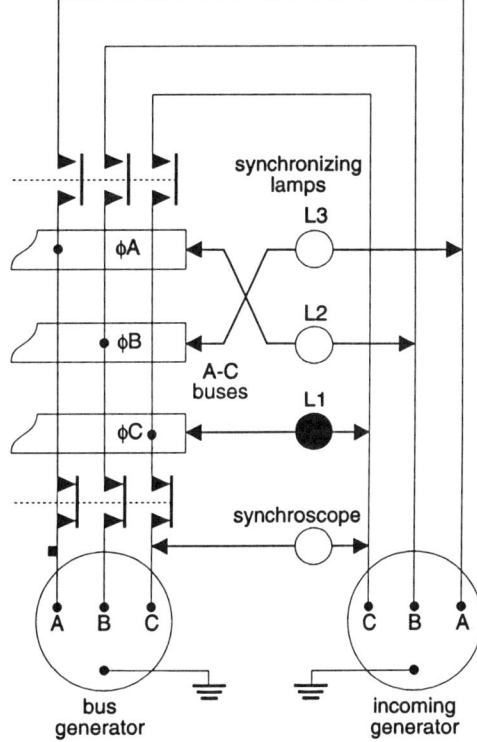

Figure 15-3. Method of paralleling a synchronous system

A sensitive synchroscope is then used as the final authority before paralleling. The synchroscope's needle freely rotates 360 degrees. The direction of rotation shows if the incoming alternator is running fast or slow. When the needle remains vertical, synchronization has been achieved.

A potential pitfall exists when independent power producers want to run their synchronous generators in parallel with the utility's power grid. In order to protect the integrity of the system, items such as voltage, frequency, and reverse power must be monitored. If any of these parameters fall outside of a specified range, that generator must be taken off-line to maintain the soundness of the system. Only approved, high-quality instrumentation is allowed to perform this function. For small installations, the relative cost of this instrumentation can be large compared with the cost of the overall system.

For safety purposes, the power company must have access to the point of interconnection with its system. Transformers don't care which way power flows through them. The step-down transformer that normally serves your facility can just as easily operate as a step-up transformer feeding high voltage into outside distribution lines. When linemen are working, they must be assured that the power lines will remain de-energized.

INDUCTION GENERATORS

An induction generator is usually nothing more than a standard, off-the-shelf induction motor driven a few percent above its rated synchronous speed. Induction generators differ from synchronous generators in that the required excitation is drawn from the power system itself; no separate dc excitation source is required.

It is possible to turn an ordinary induction motor into a generator without making any modifications to it. In order to induce current into the rotor, relative motion must be established between the rotating magnetic field and the rotor. When acting as a motor, the rotor, turning a few rpm slower than the rotating magnetic field in the stator, establishes that relative motion. When acting as a generator, the rotor is driven a few rpm faster than the stator's rotating field. This puts power back into the power network, instead of drawing power from it.

The advantages of induction generators are that they are inexpensive, their frequency control is automatic, and no elaborate precautions need to be made when paralleling.

Because the excitation is derived from the power line, voltage output and synchronizing are automatic.

Only simple instrumentation is required — usually just an ammeter to determine the generator's output. The speed of the prime mover is increased until the nameplate amperage is reached. Some people install reverse-power relays that take the generator off line in case the prime mover slows down too much. If this were to happen without a reverse-power relay installed, the generator would act as a motor, drawing power from the line and driving the prime mover (possibly causing damage to it).

The disadvantage of induction generators is that they can't be used for emergency service. Once the power to them is cut off, excitation power vanishes, and power output stops. This eliminates one of the objections utilities have against cogeneration systems. There is no danger of an induction generator feeding power back into their system during outages.

THE REFRIGERATION CYCLE

When you step out of a pool on a windy day, you can feel your skin temperature become lower. The physics behind this phenomenon is the keystone of refrigeration. When a liquid evaporates it absorbs heat, thus lowering the temperature of the surface on which it evaporates. This is roughly what happens in refrigeration and air conditioning systems. In these cases the evaporating liquid is called a refrigerant.

MECHANICAL REFRIGERATION

Let's cool a room using a refrigerant commonly used for many years, ammonia. We feed the liquid ammonia through a coil in a closed system (ammonia is toxic as well as malodorous). Liquid ammonia boils at -28°F at atmospheric pressure (14.7 psia, or pounds per square inch absolute). As long as the room's ambient temperature is warmer than -28°F, the refrigerant will absorb heat. (Remember, heat always flows from a higher temperature to a lower temperature.)

As liquid ammonia absorbs heat, some of it evaporates or boils away. During this process, the room's temperature is being lowered. That's all there is to the act of refrigeration. The ammonia gas could be vented to the outside, and the room would still be cooled. Of course, you can't afford to waste refrigerant. It is not only costly, but it is now against the law to vent refrigerants to the atmosphere due to their detrimental effect on the ozone layer. The purpose of the compressor and the rest of the equipment is to recycle refrigerant for re-use within the system.

A lot of the physics behind refrigeration technology is based on the pressure-temperature relationship of liquids. For example, water boils at 212°F at sea level. If the atmospheric pressure was reduced by climbing a mountain, water's boiling point would decrease (at 10,000 ft, the boiling point of water is 193°F). The reverse is true if the pressure is increased. If the pressure was raised to 15 psi, the boiling point of water would increase to 250°F.

Ammonia gas, on the other hand, is impossible to return to a liquid state at atmospheric pressure; at that pressure it boils at -28°F. If its pressure (and thus its corresponding boiling point) were increased enough, we could condense it back to a liquid at room temperature.

This is where the compressor fits in. If ammonia vapor is compressed to 152 psi, its boiling temperature increases from -28°F to 85°F. If the high-pressure vapor was then run through a coil exposed to, say, 75°F air, the hot vapor would give up some of its heat to the air. In the process, some of the vapor would condense to a liquid, and it would be ready to return to the evaporator to be used for cooling again. This part of the refrigeration system is appropriately called the condenser coil.

Our basic refrigeration system needs one more component. We can't use 152-psi liquid refrigerant in the evaporator; it must be reduced back down to atmospheric pressure, 14.7 psia. This is done by means of the expansion valve. Its job is to meter the proper amount of liquid ammonia into the evaporator coil. In the process, it also acts as a pressure-reducing valve.

Take a look at the schematic of a refrigeration system, Figure 15-4. This diagram is valid for refrigerators, air conditioners, and chillers. Notice that it is divided into two sections:

1. A high-pressure side that includes the compressor discharge end and condenser coil.

2. A low-pressure side that includes the evaporator coil and compressor suction end. The expansion valve straddles both sides.

ABSORPTION REFRIGERATION

Another type of refrigeration process is called an absorption system. It also recycles the refrigerant and has high- and low-pressure sides, but its operation is different.

Two fluids circulate in an absorption machine, a refrigerant and an absorbent. The two common absorption systems are lithium bromide (a salt), and ammonia. The lithium bromide system is commonly used for commercial air conditioning; it uses water as the refrigerant and lithium bromide as the

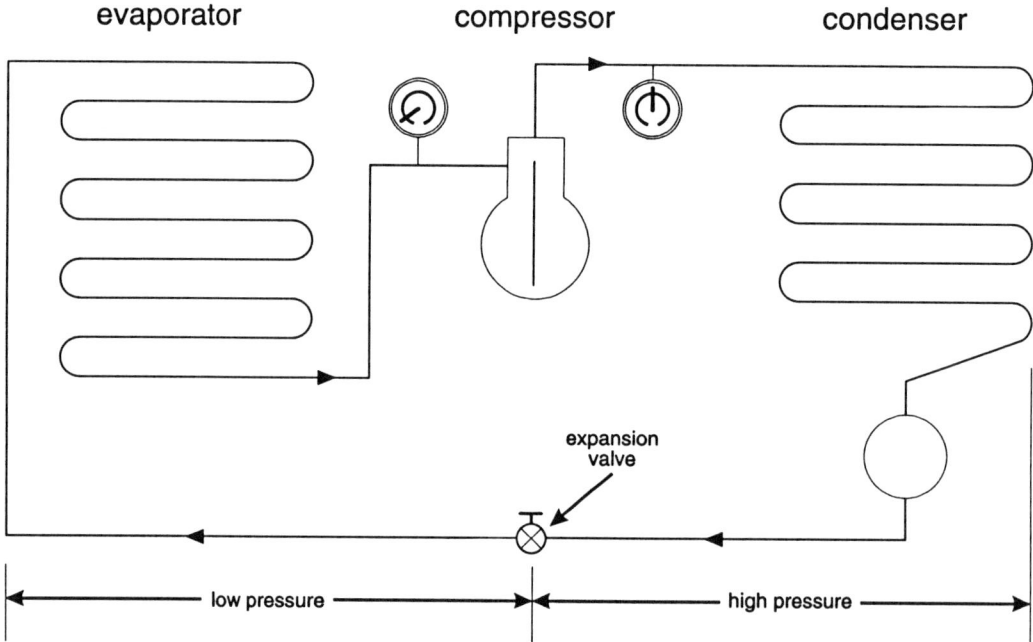

Figure 15-4. Schematic of a basic refrigeration system

absorbent. The ammonia system is capable of lower temperatures and is frequently found in industrial applications. It uses ammonia as the refrigerant and water as the absorbent.

In both of these cases, the refrigerant is strongly attracted to the absorbent. The cooler the absorbent, the more refrigerant it can hold. When heated, the absorbent gives up a great deal of the refrigerant it was holding. For the sake of brevity, this discussion will be limited to the lithium bromide system. The information is quite similar for the ammonia system.

There are four major parts to an absorber: the evaporator, absorber, generator, and condenser, Figure 15-5. Like a mechanical refrigeration system, the absorption unit is divided into a high side and low side.

The terminology may get a bit confusing, but as we outline the process, just remember that the refrigerant and absorbent take two paths. The refrigerant goes around the entire loop, while the absorbent shuttles back and forth between the absorber and generator.

Water, under very low pressure (about 0.25 in. Hg), is sprayed over the chilled-water coils in the evaporator. (Hg is the chemical symbol for mercury; in. Hg is the abbreviation for inches of mercury. Normal atmospheric pressure is around 29.9 in. Hg.) Remember, the lower the pressure, the lower the boiling point.

The chilled-water temperature leaving a lithium bromide absorber is around 42°F. Because the chilled water is warmer than the refrigerant, the refrigerant absorbs heat from the chilled water, which is cooled in

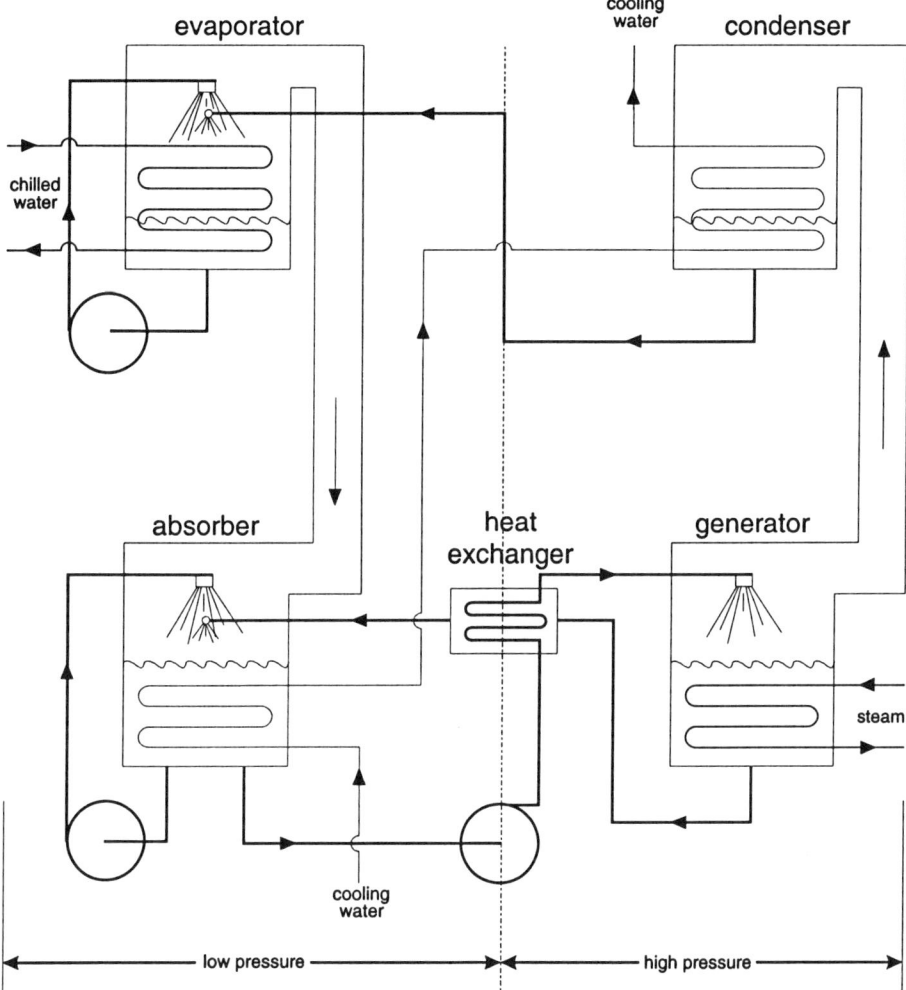

Figure 15-5. Continuous lithium bromide absorption system

the process. As the refrigerant absorbs heat, it evaporates (the same as in the mechanical refrigeration process).

The refrigerant vapor is piped to the absorber section, where it is absorbed by the concentrated lithium bromide brine. Heat is generated in this absorption process, so a water coil in the absorber section keeps the temperature down. Remember, the lower the temperature of the absorber, the more water vapor it can absorb.

Now the diluted brine is pumped over to the generator section, where heat is applied. The restriction in the line separates the low side from the high side. The high side is still under a vacuum, around 3.5 in. Hg. The water vapor is boiled out of the lithium bromide absorber and goes to the condenser.

The reconcentrated brine is returned to the absorber. The heat exchanger between the absorber and generator increases efficiency by reducing the amount of heat energy that is wasted.

The water vapor is condensed in the condenser section and returned to the evaporator, where the cycle begins again.

The heat for the generator can be supplied by steam, hot water, natural gas, or exhaust from a gas turbine or other process. The only electric power required is for the three pumps. They are usually mounted on a common shaft and driven by one motor.

Chapter 16

COGENERATION: WHAT IT IS, HOW IT WORKS, HOW TO USE IT

The most common definition of cogeneration is: the simultaneous production of electricity and usable thermal energy from a single fuel source.

A more comprehensive definition is: the simultaneous production of two or more forms of energy from a single fuel source. This opens up possibilities such as a prime-mover-driving pumps, fans, and refrigeration compressors, as well as generators. (A prime mover is a machine such as a water wheel, diesel engine, or gas turbine, that receives and modifies energy as supplied by some natural source.)

In some cogeneration system designs, natural gas or fuel oil is converted to mechanical energy by a reciprocating engine or gas turbine. The exhaust heat is then utilized as thermal energy, usually in the form of steam or hot water.

Cogeneration energy is much more efficient than electrical energy generated by large power plants. When the power company makes electricity, it rejects about two-thirds of the process-generated heat to the environment. It doesn't matter if steam turbines, gas turbines, or diesel engines are used, only one-third of the fuel burned actually ends up as "working" energy. Don't blame the power company; that's the way the laws of thermodynamics work.

With cogeneration, up to 85% of the fuel's energy can be used. Why not 100%? Temperature difference drives heat transfer. With less temperature difference, more expensive heat recovery equipment is required to utilize the lower-grade thermal energy. There comes a point when the cost of additional heat recovery equipment becomes impractical.

Figure 16-1 illustrates steam plant operation, which is basically the same concept as for gas turbine and diesel plants. In a diesel plant, for example, one-third of the fuel goes toward producing shaft horsepower, one-third goes up the stack, and the remaining third is absorbed by the cooling jacket water.

TOPPING AND BOTTOMING CYCLES

There are two general classifications of cogeneration: topping and bottoming cycles.

In a **topping cycle**, fuel is first used to produce mechanical or electrical energy, then used for other purposes. An example would be a gas turbine driving a generator, with its waste heat directed through a boiler to make steam. Most cogeneration plants use topping cycles. Figure 16-2 shows some topping cycle combinations.

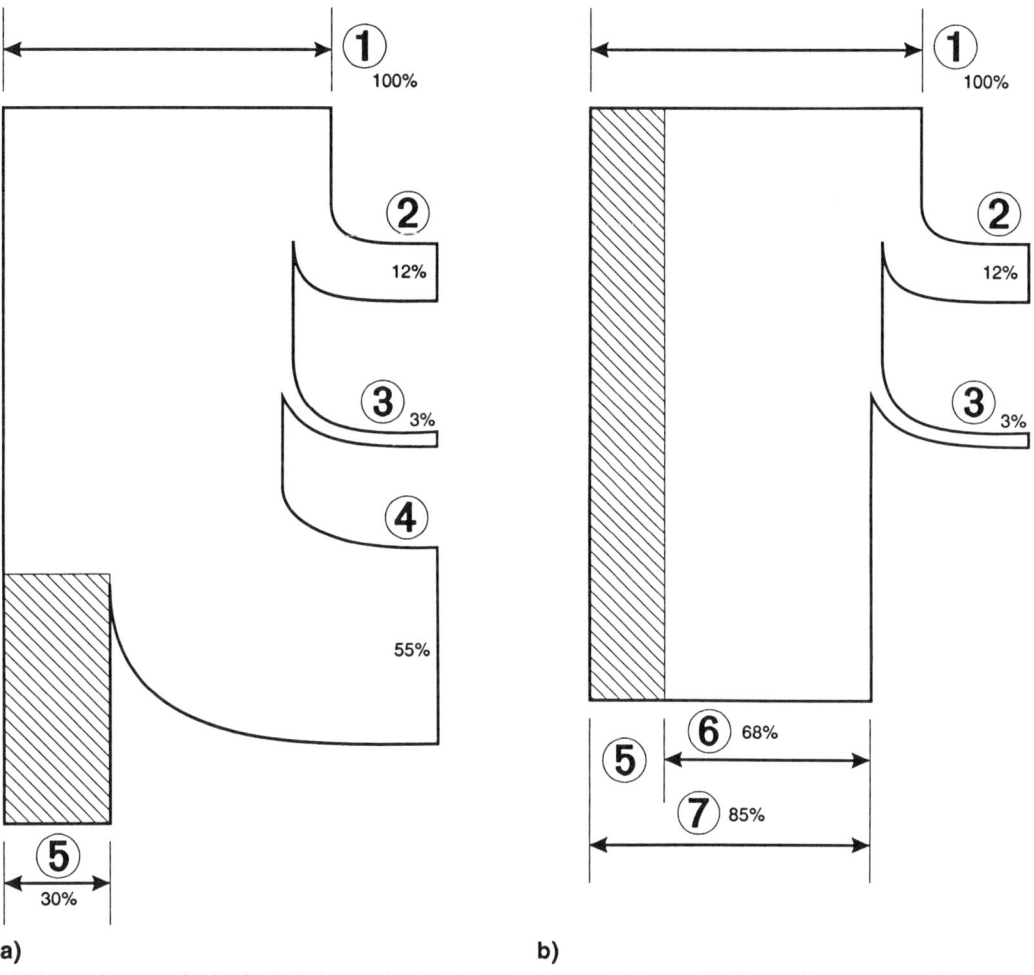

1) thermal energy in the fuel; 2) losses in the boiler; 3) losses in the auxiliaries and generator; 4) thermal energy lost to the cooling water; 5) mechanical or electrical energy produced by the turbine; 6) thermal energy from the process steam; 7) total available energy.

Figure 16-1. Illustration of basic steam plant operation; a) shows condensing turbine, b) shows back pressure turbine

In a **bottoming cycle**, fuel is first used for heating, then for other purposes. An example would be fuel first supplied to a melting furnace, whose hot flue gas is directed through a waste heat boiler. The steam could be used to drive a steam turbine or to provide process heating. Bottoming cycle plants are found in steel mills, copper refineries, and glass factories.

The point is that cogeneration is applicable for many different applications. For example, exhaust steam from a turbine can be used for additional thermal energy; and hot water is a by-product of the absorption process.

Cogeneration: What it is, How it Works, How to Use it

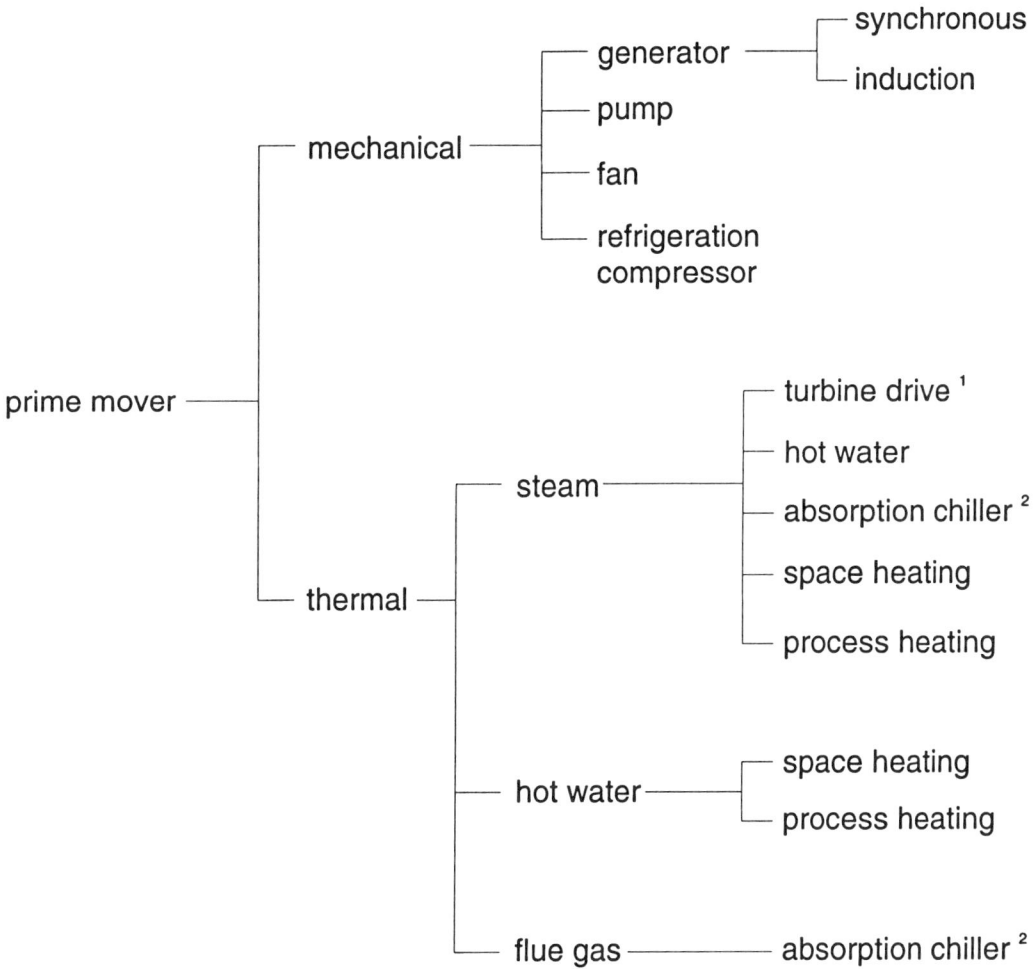

[1] Exhaust steam from turbine can be used for additional thermal energy.
[2] Hot water is a by-product of the absorption process.

Figure 16-2. Topping cycle combinations

DON'T WASTE HEAT

The economics of cogeneration depends on finding a use for waste heat. When heat rejected from the prime mover can't be used, it is cheaper to buy power from the utility. If the waste heat can be put to good use, however, then a detailed economic and engineering evaluation is justified.

The more often a cogeneration facility is operated during the year, the faster the capital investment is paid back. Hours of operation weigh heavily in the appraisal, and often makes or breaks the economics of a project.

As you may have suspected, cogeneration comes with a relatively high price tag. First there is the initial capital investment. Then there are operation, maintenance, and fuel expenses.

When evaluating cogeneration system possibilities, make sure there is enough commitment in your organization to see the thing through for the next 20 years. Budget enough money for operation, training, and maintenance.

Some cogeneration projects are modest and require a minimum of supervision. The simpler the better! But let's say one of the proposals is an ambitious, operate around-the-clock utility-type operation. If you aren't accustomed to managing this type of endeavor, keep in mind that it takes a minimum of four people working on a rotating basis to cover the shifts.

Even with four people, there is always one shift a week that will be left open or require overtime. Three shifts per day, times seven days a week, equals 21 shifts per week. Five shifts per week per person, times four persons, equals 20 shifts per week. Many operations employ a "fifth shift" that fills in on open days and vacations.

DOWN-SCALED PLANS

If there isn't the inclination, wherewithal, or expertise for this magnitude of undertaking, but the numbers look good and you still want to enjoy the benefit of cogeneration, there are alternatives. One of the most popular is a third-party arrangement. These outfits will finance, design, build, and operate the installation.

Third-party companies are in business to earn a profit. They sell the electrical and thermal energy produced in their own facilities. Typically, you or a group of users agree to purchase the thermal output of the plant, at less than the present costs. Sometimes an existing boiler plant can be retired by making such an arrangement.

There could also be an agreement to purchase all or part of the plant's electrical output. Excess electrical capacity is usually sold to the local utility. In the meantime, you still get a break from electric and thermal energy rates, but don't have the hassle of operating the plant. And these third-party companies specialize in power plants; they have more expertise and resources than the average organization.

With cogeneration, the heat rejected by a prime-mover system is used to power a process. If a diesel engine is driving a load, the heat normally rejected from the exhaust and water jackets can be used to make steam. The heat from the oil cooler also could be used to make hot water.

This waste heat is most commonly used to heat water, which changes state and becomes steam; steam drives the power generator. Common practice states that if low-pressure steam is required, a diesel engine should be considered. (Low-pressure steam is defined as 15 psi or under.) If high-pressure steam is required, a gas turbine would be a better choice. The hot exhaust would be directed to a waste heat boiler, sometimes referred to as a heat-recovery steam generator (HRSG).

If your business requires large quantities of hot water to power its processes, you definitely should take a look into cogeneration. Not only can the exhaust heat be recovered, but the heat from the jacket cooling water can be, too.

If you think your organization has no uses for process steam, just consider a steam-driven absorption chiller for air conditioning. (Yes, hot steam can be used to power a cooling system.)

STEAM-POWERED ABSORPTION COOLING

Absorption cooling systems have definite benefits — especially in applications with a plentiful supply of excess steam. For starters, they don't use CFC-based refrigerants.

CFCs — chlorofluorocarbons — are getting a lot of press for depleting the ozone layer. Absorption chillers use pure H_2O as the refrigerant.

One of the fairly recent innovations for small installations has been in the air conditioning field. A natural gas-fueled industrial engine can drive a refrigeration compressor. In addition, the rejected heat from the engine is being used in an absorption chiller to produce additional cooling.

Even though no electricity is being generated, none is being consumed, either. Electric-driven chillers are cheaper to run than absorption machines when only energy charges are used. When electric demand charges are figured in, it's possible that an absorption system would be more economical. And, if steam produced from an in-house process powers the absorption system, the payback looks more attractive.

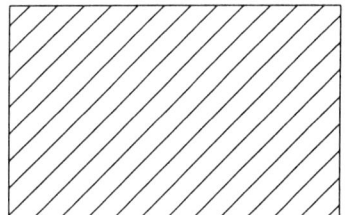 # GLOSSARY

A

Ac (Alternating Current) — An electric current that constantly changes magnitude and periodically changes direction. The frequency of the alternations are measured in hertz (cycles per second).

Algorithm — A set of well-defined rules or procedures for solving a problem or providing an output from a specific set of inputs.

Alternator — An ac generator.

Ammeter — An instrument for measuring electric current.

Ampacity — The current in amperes a conductor can carry continuously under the conditions of use, without exceeding its temperature rating.

Ampere (Amp) — A unit of electrical current or rate of flow of electrons; 1 V across 1 ohm of resistance causes a current flow of 1 amp. A flow of 1 coulomb per sec equals 1 amp.

Analog Input — A continuously variable electrical input that travels from a sensing device that reflects incremental changes in voltage, current, temperature, humidity, light level, pressure, or other parameters.

Annunciator — A signaling apparatus which, by visual or audible means, indicates whether a voltage or current is or has been present in one or more circuits. Usually associated with alarm circuits; alerts operators of abnormal conditions.

Armature — Windings in motors and generators: In the case of the former, used where power is received; and in the case of the latter, used where power is delivered. In dc machines the armature usually rotates; in ac machines the armature is usually stationary.

Armored Cable (Type Ac) — Trade name is "BX" cable; a fabricated assembly of insulated conductor in a flexible metallic enclosure. Metallic enclosure is permitted to act as the grounding conductor.

ASCII (American Standard Code for Information Interchange) — An 8-bit coded character set to be used for the general interchange of data among information-processing systems, communications systems, process control systems, and associated equipment.

Attenuation — Also called suppression or rejection; the amount of signal loss in a system. In a device such as an isolation transformer, the degree of reduction of unwanted spikes and signals, usually expressed as a voltage ratio or in decibels.

Apparent Power — The product of voltage and current in a circuit, in which the two may or may not reach their peaks at the same time. Units are expressed in VA or kVA. The distance between peaks is expressed as a phase angle.

Average Value — The sum of a quantity of peak instantaneous values in a half cycle, divided by the quantity of instantaneous values. In a sine wave, average value = 0.637 x peak values.

Autotransformer — A transformer in which part of the winding is common to both the primary and secondary circuits.

AWG (American Wire Gauge) — A system of numerical designations for small solid and stranded conductors used in homes, industry, and electric utilities.

B

Balanced Load — An ac power system using more than two wires (i.e. single-phase, three-wire; three-phase, three-wire; and three-phase, four-wire), where the current load-carrying lines all carry the same current.

Bandwidth — The ability of a device to respond to rapid changes in a measured parameter without overshoot or undue delay.

BAUD — A unit of signaling speed equal to the number of discrete conditions or signal events per sec.

Bayonet — A lampholder for low-voltage incandescent lamps, with an unthreaded metal shell with two diametrically opposite keyways that cooperate with similarly located projections on a mating lamp bulb. Pushing down on the bulb and turning it clockwise in the socket locks the bulb in place.

BIL (Basic Impulse Level) — The insulation class of a device, expressed as the impulse crest value of withstand voltage of a specific full-impulse voltage wave.

Billing Demand — The demand level a utility uses to calculate demand charges on an electric bill.

Bonding — The permanent joining of metallic parts, to form an electrically conductive path that ensures electrical continuity and the capacity to safely conduct any current likely to be imposed.

Boost — Single- or three-phase autotransformers connected to increase voltage in small amounts.

Branch Circuit — That portion of a wiring system extending beyond the final overload protective device.

Breakdown — A disruptive discharge through insulation.

Buck — Single- or three-phase autotransformers connected to decrease voltage in small amounts.

Bus — A conductor or group of conductors that serves as a common connection for three or more circuits in a switchgear assembly.

Burden — The VA load an instrument places on a current or potential transformer. If this burden exceeds the transformer's rated capacity, accuracy deteriorates.

Bushing — A lining for a hole for insulation and/or protection from abrasion of the conductors passing through it.

"BX" Cable — See "Armored Cable."

C

Cable — A preassembled bundle of conductors, insulating material, and sheathing that make up a ready-to-install cabling system.

Cable Sheath — The protective covering, such as lead or plastic, applied over a cable.

Candelabra — A small screw-base threaded lampholder accepting a bulb approximately ½ in. dia, commonly used in night lights, indicator lights, and Christmas tree bulbs.

Capacitor — A device essentially consisting of two conducting surfaces separated by an insulating material or dielectric, such as air, paper, mica, plastic film, or oil. A capacitor stores electrical energy, blocks dc flow, and permits ac flow to a degree dependent on the capacitance and frequency. Units are measured in farads, microfarads, or micro-microfarads.

Circuit Breaker — A device designed to open and close a circuit by non-automatic means, and to open the circuit automatically on a predetermined overcurrent without injury to itself when properly applied within its rating.

Common Mode Noise — Signals or spikes impressed from line-to-ground in a power distribution system.

Conductor — A material through which an electric current can flow relatively easily.

Conduit — A raceway of circular form for enclosing and protecting wires or cables. Can be rigid or flexible, metal or plastic.

Connected Load — The total load a customer can impose on the electrical system if everything was connected simultaneously. Connected load can be measured in hp, W, or VA.

Contract Load Capacity — A load capacity contracted with a utility and related to connected load. The minimum monthly demand charge is sometimes established by applying the demand rate to some specified percentage of the contracted capacity.

Converter — A device that changes ac to dc.

Crest Factor — The ratio of the crest, peak, or maximum value in a periodic function to its root mean square (RMS) value.

CT (Current Transformer) — Used in an ac metering circuit, it matches line current to an instrument's range. Caution: An energized CT with a disconnected secondary is a shock hazard that can also cause equipment damage.

CT Ratio (Current Transformer Ratio) — The ratio of primary amps divided by secondary amps.

D

Dead Front — Without live parts exposed to a person on the operating side of the equipment.

Dc (Direct Current) — An electric current that flows steadily in one direction.

Decibel (dB, bel) — A logarithmic unit. The fundamental logarithmic unit is the bel, which is the common logarithm of the ratio of two values of power. The practical unit is the decibel (dB), which is 0.1% of a bel. The decibel may also be used to express voltage or current ratios.

Delta Winding — A method of connecting the windings of a three-phase transformer or motor, so the windings form an electrical triangle whose corners are the connections of the three phases.

Demand — A measure of the customer load connected to the electrical power system at any given time. Units are usually measured in kW or kVA.

Demand Charge — The charge utilities apply to the billing demand for the current month. Units are usually measured $/kW or $/VA.

Demand Interval — A specific interval of time on which a demand measurement (kW, kVA, kVAR, etc.) is based. Common intervals are 15, 30, and 60 min.

Demand Penalty — A financial penalty levied by an electric utility on its customer for exceeding a mutually established kW, kVA, or kVAR level.

Discretionary Loads — Loads that exhibit a "flywheel" effect, so that removing them from the line for short periods does not affect operation or comfort. Typical are hvac systems, hot water heaters, and snow-melting systems.

Double Pole, Double Throw (DPDT) — A switch that makes or breaks the connection of two conductors to two separate circuits.

Double Pole, Single Throw (DPST) — A switch that makes or breaks the connection of two circuit conductors in a single branch circuit.

Duct — In electricity, a single enclosed runway for conductors or cables.

Duplex Outlet — Two receptacles in a common housing or mounting that accepts two plugs.

E

Eddy Current — Current flowing in the windings of a transformer (in addition to the load current), caused by the magnetic field flux and producing heating loss in the conductors (in addition to the load heating loss).

Edison Base — A lampholder with a threaded internal shell (approximately 1 in. dia) that accepts lamp bulbs of the size commonly used for domestic illumination.

Effective Value — See "RMS."

Efficiency — The ratio of the output power divided by the input power. Usually expressed as a percentage.

Electrostatic Shield — Typically, one turn of a thin sheet of aluminum or copper extending over the full width of the windings of a transformer, usually located between the primary and secondary windings.

EMT (Electrical Metal Tubing) — A thin-walled steel raceway of circular form for the protection of wire or cables; thin-wall conduit.

Energy Consumption Charges — The charges a utility imposes for the consumption of real power in watts. Units of measurement are usually $/kWh.

Energy (Electrical) — The ability of electric power to do work; Energy = (power) x (time); 1 Joule of energy = 1 W-sec.

EPO (Emergency Power Off) — A switch installed in a strategic location, usually by an exit, activated by pushing a red mushroom-shaped handle; it shuts off all power, except lights, in the room. Commonly found in computer centers.

Exciter — An auxiliary dc generator used to supply power to the field of a synchronous generator or motor.

F

Fault — A local failure in the insulation on a conductor.

Fault Current — The abnormal current flowing between conductors or from a conductor to ground due to an insulation defect, arc-over, or incorrect connection.

Feeder — All circuit conductors between the service equipment and the final branch-circuit, over-current device.

Field Windings — In the case of motors and generators, the windings that provide the magnetic field. In dc machines the field is usually stationary; in ac machines the field usually rotates.

Filter — Capacitors and inductors combined to provide a reactance that is selective, or "tuned," to a harmonic frequency in order to reduce its magnitude. Shunt filter across a load provides a low-impedance path to bypass the harmonic; series filter (in series with a load) provides a high impedance to block the harmonic.

Four-Way Switch — A switch installed between pairs of three-way switches to control one electrical load from three locations.

Frequency — A measure of cycles of a wave motion in a time period; the number of recurrences of a periodic event per period of time. In electrical terms, frequency is specified as so many Hertz (Hz) or cycles per sec (cps).

Frequency, Fundamental — The basic or lowest frequency of a wave form.

Fuel-Adjustment Charges — The charges a utility imposes for changes in the cost of the fuel they use and other utility cost factors. Frequently, these charges are based on complex formulas that include many variables related to the cost of delivering electrical energy. Units are usually $/kWh.

Fuse — A protective device with a fusible element that opens the circuit by melting when subjected to excessive current.

G

GFCI (Ground Fault Circuit Interrupter) — A receptacle integral with a circuit-interrupting device that detects leakage current to ground on the load side, by activating a circuit-interrupting device. Reduces electric shock hazard by instantly opening a circuit when detecting current flowing to ground instead of through the neutral conductor.

GFI (Ground Fault Interrupter) — Similar to a GFCI; when it detects leakage current to ground, it opens a circuit breaker. Usually installed at the service entrance, it's primarily used to protect equipment instead of people.

Greenfield — Flexible metal conduit made with interlocking-edge, helically wound steel.

Ground — A connection, intentional or accidental, between an electrical circuit or equipment and the earth, or to some conducting body that serves in place of the earth.

H

Harmonic — A sinusoidal component of a periodic wave or quantity, having a frequency that is an integral multiple of the fundamental frequency. This is often superimposed on and thus distorts the fundamental current or voltage sinusoidal wave shape.

Harmonic Analyzer — An electronic device for measuring the amplitude and phase of various harmonic components of a wave form.

Harmonic Distortion — Non-linear distortion of a system; characterized by the appearance of harmonics other than the fundamental component, when other than the input wave is sinusoidal. The ratio of the total RMS value of all harmonics to the RMS value of the fundamental; when expressed as a percent, it's called Total Harmonic Distortion. Variable-frequency motor controls, computer power supplies, and other SCR and triac switching devices distort the current in a circuit. Excessive harmonics cause overheating in neutrals and transformers.

Hertz (Hz) — A unit of frequency; 1 Hz = 1 cycle per sec.

Hicky — A fitting for mounting a lighting fixture in an outlet box. Also, a device used with a pipe handle for bending conduit.

High Voltage — Various agencies define high voltage differently. For utilities it is between 15 and 250 kV.

Horsepower (Hp) — A unit of power; 1 hp = 746 W = 33,000 lb-ft/min.

Hospital-Grade Equipment — A connector, plug, receptacle, or other equipment designed to meet additional performance requirements of high-abuse areas often found in hospital locations. Such connectors are tested to "Hospital-Grade" requirements of UL Standard 498.

I

Impedance (Z) — The total resistance and reactance opposition a circuit offers the flow of alternating current. It is measured in ohms.

Induced Current — A current that results in a closed conductor due to cutting lines of magnetic force.

Inverter — A device that changes dc to ac, or one frequency of ac to another.

Induction Motor — An ac motor in which the primary winding (usually the stator) is connected to the power source and induces a current into a polyphase secondary (usually the rotor).

Inductor — A device consisting of one or more associated windings, with or without a magnetic core, for introducing inductance into a circuit. Transformers, lamp ballasts, and motors are inductors.

Initiator Pulses — Electrical impulses generated by pulse-initiator mechanisms installed in utility revenue meters. Each pulse represents the consumption of a specific amount of watts. The pulses are used to measure energy consumption and demand.

Inverter — A device that changes dc to ac. Found in UPS and variable-speed drive machines.

IR Drop (Voltage Drop) — A potential difference (voltage) produced by current flowing through a resistance; the phenomenon that reduces voltage at the end of a power line.

I^2R Losses — A variation of the basic power equation; often used to describe the power consumed, and therefore lost, in power lines.

Isolated Ground — A grounding-type receptacle in which the equipment ground contact and terminal are electrically isolated from the receptacle mounting device. It can be used to prevent ground loop currents from interfering with sensitive equipment.

Isolation Transformer — Isolates a load from a dc voltage or from ground; any transformer with windings electrically isolated from each other. Isolation transformers may or may not have electrostatic shields.

K

Kcmil — One-thousand circular mills; a unit of measure for large-size distribution and feeder cables (replacement for MCM).

Kilowatt (kW) — 1,000 W.

Kilowatt-Hour (kWh) — A unit of electrical measurement indicating the expenditure of 1,000 W for one hour; a measure of energy transfer. Higher quantities are expressed in megawatt-hours (MWh), or the expenditure of 1 million W for one hour.

kVA — Kilovolt-amperes.

L

Lagging Current — The current flowing in a circuit that is mostly inductive. If the circuit contains only inductance, current lags the applied voltage by 90 degrees.

Leading Current — The current flowing in a circuit that is mostly capacitive. If the circuit contains only capacitance, the current leads the applied voltage by 90 degrees.

Lightning Arrester — A device that reduces the voltage of a surge applied to its terminals and restores itself to its original operating condition.

Linear Load — Where current is directly proportional to the applied voltage.

Load — Any device or circuit that consumes power in an electrical system.

Load Restoring — The energizing of loads that were previously removed from the line to limit load and control demand level.

Load Shedding — The removal of loads from the line to limit load and control demand level.

Low Voltage — Different agencies define low voltage differently. For utilities it is below 600 V.

M

Maintained Contact — A switch that, when the actuator is moved to the "on" position, makes and retains circuit contact until the actuator is manually moved to the "off" position.

MCM — One-thousand circular mills; a unit of measure of large-size distribution and feeder cables. The National Electrical Code replaced this term with kilo circular mills (kcmil).

Medium Voltage — Different agencies define medium voltage differently. For utilities it is between 600 V and 15 kV.

Mercury — A type of switch construction employing liquid mercury as the contact means for making and breaking an electrical circuit.

Metal-Clad Cable (Type-MC Cable) — A factory-assembly of one or more conductors, each individually insulated and enclosed in a flexible metallic sheath of interlocking tape, or a smooth or corrugated tube.

Mho — A unit of conductance; the reciprocal of resistance (ohm spelled backwards). The unit "Siemens" is replacing the mho. Working units are micro-mhos and micro-Siemens.

Modem — Acronym for modulate-demodulate; used to send and receive computer data over telephone lines.

Momentary Contact — A normally open switch that establishes circuit contact when its actuator is moved to, and held, in the "on" position. The circuit is broken when the actuator is allowed to return, by itself, to the "off" position. It also may be furnished to operate in the normally closed mode.

N

N.C. — Abbreviation for a normally closed switch or contact.

Neutral — The conductor chosen as the return path for the current from the load to the source of power. The neutral is frequently, but not necessarily, grounded.

N.O. — Abbreviation for a normally open switch or contact.

Non-Linear Load — A load in which the current is not proportional to the applied voltage.

Nonmetallic Sheathed Cable (Type NM Cable) — Trade name "Romex"; a factory-assembly of two or more insulated conductors having an outer sheath of moisture-resistant, flame-retardant, nonmetallic material.

O

Off-Peak Power — Power supplied during designated periods of relatively low system demands, generally during the evening and early-morning hours.

Ohm — The unit of electrical resistance; 1 ohm is the value of resistance through which a potential difference of 1 V maintains a current flow of 1 amp.

Ohmmeter — An instrument for measuring resistance values.

Ohm's Law — The voltage across an element of a direct current circuit is equal to the current in amperes through the element, multiplied by the resistance of the element in ohms.

Open-Circuit Voltage — The terminal voltage of a conductor under conditions of no current demand. Since there is no current flow, there is no IR or voltage drop; thus, terminal voltage equals supply voltage.

On-Peak Power — Power supplied during designated periods of relatively high system demands, generally during the working day.

Oscilloscope — An instrument for displaying the wave forms of ac voltages.

Outlet — A point in the wiring system from which current is taken for the supply of lamps, appliances, etc.

P

Peak-Average Demand — The highest average load over a utility-specified interval during a billing period. If there is no ratchet clause in the rate schedule, the peak-average demand is also the billing demand.

Peak-to-Peak Value — The amplitude of an ac wave form from its positive peak to its negative peak.

Phase Angle — The difference in degrees by which current leads voltage in an inductive circuit.

Plug — A device with male contacts which, when inserted into a receptacle, establishes connection between the conductors of the attached flexible cord and the conductors connected to the receptacle.

Polyphase — Having or utilizing several phases; a polyphase ac power circuit has several phases of ac with a fixed phase angle between phases.

Polarity — The electrical characteristic of voltage that determines the direction of current flow.

Potential Difference (Voltage) — The force that allows current flow.

Potential Transformer (PT) — An instrument transformer, the primary of which is connected in parallel with the circuit whose voltage is to be measured. Used to step down high voltages to lower levels (120 V) acceptable to measuring instruments.

Potentiometer — A resistor with a continuously variable contact arm, used to change voltage to a load from a voltage source. Similar to a rheostat; however, electrical connections are made to both ends of the resistor and to the arm.

Power — The rate of doing work. When 1 V supplies 1 amp to a load, power equals 1 W.

Power Factor (PF) — True power in watts divided by the apparent power in volt-amperes. Also, the cosine of the phase angle between the voltage applied to a load and the current passing through it.

Power Factor Correction — Steps taken to raise the power factor by bringing the current more nearly in phase with the applied voltage. Most frequently, this consists of adding capacitance to increase the lagging power factor of inductive circuits.

Power Factor Penalty — The charge utilities impose for operating at power factors below some specified level.

Primary — The input winding of a transformer.

R

Raceway — A channel for holding wires or cables that can be constructed from metal, wood, or plastics. Raceways can be found in the form of conduit, tubing, surface-mounted types, or cast-in-place, under-floor types.

Ratchet Clause — A rate schedule clause stating that billing demand may be based on current month peak average demand or historical peak average demand, depending on relative magnitude. Billing demand is either the current month's peak average demand or some percentage of the highest historical peak average demand, depending on which is largest.

Rating — The rating of a device, apparatus, or machine that sets the limits of its operating characteristics. They are commonly stated in volts, amps, watts, degrees, horsepower, etc.

Reactance — Opposition to the flow of alternating current. Capacitive reactance is the opposition offered by capacitors; inductive reactance is the opposition offered by an inductive load. Both are measured in ohms.

Reactive Power (Volt-Amperes Reactive; VAR) — Also called wattless power, reactive power increases with decreasing power factor. It is the component of apparent power which does no real work.

Real Power (Watts) — The component of apparent power that represents true work, as in the ac circuit. It equals apparent power times the power factor.

Receptacle — A device with female contacts, primarily installed at a structure or in a piece of equipment, which is intended to establish electrical connection with an inserted plug.

Rectifier — A device (such as a diode) that changes ac to dc. It allows current flow in only one direction, during only part of an ac voltage cycle.

Resistance — The property of a substance that impedes current flow and results in the dissipation of power in the form of heat. The unit of resistance is the ohm; 1-ohm resistance is the resistance through which a difference of the potential of 1 V produces a current of 1 amp.

Rheostat — A variable-resistive device consisting of a resistance element and a continuously adjustable contact arm. Similar to a potentiometer; however, connections are only made to one end of the resistance element and the adjustable contact arm.

Resonant Frequency — The frequency of a circuit containing capacitance and inductance, at which capacitive reactance equals inductive reactance, and at which the circuit tends to oscillate.

Root Mean Square (RMS) — The effective dc heating or I^2R value of ac; used for an accurate computation of power in watts. The RMS heating value is the same as if continuous dc of the same voltage was applied to a pure resistance. (RMS = 0.707 of peak value for a sine wave.)

Romex Cable — See "Nonmetallic-Sheathed Cable."

S

SCADA (Supervisory Control and Data Acquisition) — A system of computers and data multiplexers used to gather data and control processes.

SCR (Silicon-Controlled Rectifier) — A solid-state device that conducts current above a particular applied voltage.

Scott Connection — Connections for transformers to change two-phase power to three-phase power, and vise versa. (Also called a T-connection.)

Secondary — The output winding of a transformer.

Series Circuit — A circuit that provides a complete path for any current, that has its components connected end-to-end. If any of the components are disconnected, the circuit is broken.

Service Equipment — Usually consisting of a circuit breaker or switch and fuses, and their accessories, located near the point of entrance for the supply conductors to a building or other structure. It is intended to constitute the main control and means of power supply cutoff.

Short-Circuit — A fault path, usually unintentional, that conducts excessive current and has the potential for damage.

Shunt — A parallel connection.

Single-Phase Power — An ac circuit in which only one phase of current is available in a two- or three-conductor system, where the load lines are 0 to 180 degrees out-of-phase.

Single Pole, Double Throw (SPDT) — A switch that makes or breaks the connection of a single conductor with either of two other conductors.

Single Pole, Single Throw (SPST) — A switch that makes or breaks the connection of a single conductor in a single branch circuit.

Sinusoidal Voltage and Current — Effective value = 0.707 x peak value = RMS value; Average value = 0.637 x peak value; Peak value = 1.57 x average value; Average value = 0.9 x effective value.

Sliding Demand Interval — A method of calculating average demand by averaging the demand over several successive short intervals, advancing one short interval each time. Updating average demand at short intervals gives the utility a better measure of the true demand and makes it difficult for the customer to conceal high, short-term loads.

Slip — In an induction motor, the difference in rpm between the rotating magnetic field and the shaft; often expressed as a percentage.

Snake — A steel wire or flat ribbon with a hook on one end; used to pull wires through conduit.

Solenoid — An electric coil that, when energized, pulls in or breaks electrical contacts; a term used to describe a heavy-duty relay.

Star Connection — See "Wye Connection."

Static Disconnect — An isolation switch placed on both sides of a power circuit breaker or transformer, used to isolate equipment from all power sources for maintenance or replacement.

Submersible Apparatus — Apparatus constructed so that it will operate under water at predetermined conditions of pressure and time.

Substation — A cluster of equipment containing switches, circuit breakers, buses, and transformers for switching power circuits and transforming power from one voltage level to another, or from one system to another.

Surface-Mounted — A receptacle or other device intended to be installed on the surface of the structure or equipment in which it is mounted.

Surge — A transient variation in current and voltage at a given point in a circuit.

Switch — A device for making, breaking, or changing the connections in an electrical circuit.

T

T-Connection — See "Scott Connection."

Three-Phase Power — A system using three conductors, whose sine waves are displaced 120 degrees from each other.

Three-Way Switch — A switch used in pairs to control one electrical load from two locations.

Thyristor — A solid-state rectifier that conducts current after a firing pulse is applied.

Time-of-Day Metering — A utility program intended to offer incentives for using power during off-peak hours.

Transducer — A device that receives information in the form of one physical quantity, and converts it to information in the form of the same or another physical quantity.

Transformer — A device that steps voltage or current up or down for maximum power transfer, electrical isolation between primary and secondary, and, in special designs, for automatic regulation of voltage or current.

Transformer Loss — The sum of the no-load losses in excitation windings, plus all others from heating, including eddy current and other stray losses.

Transient — A deviation from the ideal ac sine wave of short duration compared to one cycle of the sine wave (may be non-periodic); any disturbance in the normal distribution power wave, from long-term voltage dips and rises of several millisec, to spikes and signals of less than 1 microsec duration. Typical sources of transients and noise are switching surges on large distribution systems, lightning, arcing contacts on relays and conductors, welding equipment, motor brushes, and SCRs.

Transverse Mode Noise (Differential Mode Noise) — Signals or spikes impressed from line to line in a power-distribution system.

Triplen Harmonics — Odd harmonics that are multiples of three; they increase ac current in the common or neutral conductor of three-phase, four-wire systems.

True Power — See "Real Power."

U

Unbalanced Loads — An ac power system using more than two wires (i.e. single-phase, three-wire; three-phase, three-wire; three-phase, four-wire), where current is not equal in the current-carrying wires due to uneven loading of the phases.

UPS (Uninterruptible Power Supply) — Provides ac power when normal power fails; main components are a rectifier, batteries, and inverter. Usually sized to supply power for 15 to 30 min; during this time an orderly shut down is made, or in-house power is brought on line.

V

VA (Volt-Ampere) — A unit of apparent power that equals volts times amperes, regardless of power factor.

VAR (Volt-Ampere Reactive) — A unit of reactive power as opposed to real power in watts. One VAR is equal to one reactive volt-ampere.

Vibrator — An obsolete electromechanical device that changed dc into pulsating dc (simulated ac), so it could be fed to a transformer and transformed to a higher dc voltage after being rectified. Found in car radios until the mid-50s; vibrators have been replaced by solid-state devices.

Volt-Ampere Demand — Where peak average demand is measured in volt-amperes rather than watts; in this case, the customer is automatically penalized for operating at low power factor.

Volt (V) — The force that causes electrical current to flow through a conductor; 1 V equals the force required to produce a current flow of 1 amp through a resistance.

Volt-Ampere Reactive (VAR) Demand — Measuring VAR demand is a method of penalizing for poor power factor. Multiplying total peak average VAR demand by some rate ($/VAR) penalizes for operating at any power factor less than unity.

W

Watt (W) — A unit of the electrical power required to "work" at the rate of 1 joule per sec; 1 W = 44.26 ft-lb/min = 10^7 ergs/sec = 0.00134 hp. It is the power expended when 1 amp of dc flows through a resistance load of 1 ohm. Watts are called true power.

Watt Demand — The usual demand billing factor, where peak average demand is measured in watts or real power.

Watt-Hour (Wh) — A unit of electrical work indicating the expenditure of 1 W of electrical power for one hour.

Wattmeter — An instrument for measuring the real power in an electric circuit. Its scale is usually graduated in kW.

Wiper — An electrical contact arm.

Work — The product of force by distance through which the force acts. Work is numerically equal to energy.

Wye Connection — A type of three-phase transformer or motor configuration with all phase windings connected to a common point.